AF594985

Adaptation of the Root System to the Environment

Adaptation of the Root System to the Environment

Editor

Antonio Montagnoli

MDPI • Basel • Beijing • Wuhan • Barcelona • Belgrade • Manchester • Tokyo • Cluj • Tianjin

Editor
Antonio Montagnoli
University of Insubria
Italy

Editorial Office
MDPI
St. Alban-Anlage 66
4052 Basel, Switzerland

This is a reprint of articles from the Special Issue published online in the open access journal *Forests* (ISSN 1999-4907) (available at: https://www.mdpi.com/journal/forests/special_issues/Adaptation_Root_Environment).

For citation purposes, cite each article independently as indicated on the article page online and as indicated below:

LastName, A.A.; LastName, B.B.; LastName, C.C. Article Title. *Journal Name* **Year**, *Volume Number*, Page Range.

ISBN 978-3-0365-3969-0 (Hbk)
ISBN 978-3-0365-3970-6 (PDF)

Cover image courtesy of Antonio Montagnoli

Contents

About the Editor

Antonio Montagnoli is an associate professor of Environmental and Applied Botany at the Department of Biotechnology and Life Science at the University of Insubria (Varese, Italy). His work is dedicated to the plant morpho-physiological responses to environmental and anthropogenic drivers, such as soil erosion, drought, fire, mechanical constraint, and forest management. In general, he is interested in plant root development from a wide perspective. In recent years, he has focused his attention on the traits of both coarse and fine roots. In particular, he studies the coarse roots' spatial deployment as a function of plant stability and fine root dynamics as a function of water and nutrient adsorption strategy. His work covers both the seedling and mature tree growth stages as well as the single-tree or community levels. More recently, different topics such as LED-light influence on plant growth, plant phenotyping, root-soil strengthening, Lidar data for root and forest biomass estimation, biochar influence on root dynamics, and genes, hormones, and proteins involved in lateral root formation have been under investigation. To date, he is the author of more than 40 papers published in peer-review international journals.

MDPI

Editorial

Adaptation of the Root System to the Environment

Antonio Montagnoli

Department of Biotechnology and Life Science, University of Insubria, Via Dunant, 3, 21100 Varese, Italy; antonio.montagnoli@uninsubria.it

The plant fine roots system (i.e., diameter smaller than 2 mm) plays a fundamental role in water and nutrient uptake [1], while bigger root fractions, such as large and coarse roots, ensure plant anchorage and stability [2]. Both individual roots and root spatial placement promote effective anchorage of trees [3,4]. Root systems fulfill these functions by responding to alterations in environmental conditions through a series of changes at the morphological, physiological, and molecular levels [5–8]. The rooting environment is influenced by natural conditions (resource availability, sloppy terrain and prevailing high wind) [9–12] as well as human activities (logging, fire, etc.) [13–15]. In response to external disturbances, the finest fraction of the root system modifies traits such as length, diameter, specific root length, and life span [9], whereas the coarse root fraction is spatially displaced to accomplish maximum plant anchorage [10]. Furthermore, wood production in coarse roots is laid down asymmetrically, resulting in an eccentric pattern [3,4].

In the paper collection presented here, we aimed at putting together different manuscripts that illustrate the relationship between plant roots and the environment from a wide perspective and at all levels of investigation (seedling, single tree and community, ranging from morphology to molecular biology). We also looked at innovative research in terms of new technologies that boost the discovery of new insights in root ecophysiology and biology. It is important to highlight that the present collection has taken steps in parallel timing with another Special Issue [16], which was more specifically devoted to forest ecosystems. Together, these two collections clearly demonstrated the vivid activity of the root science community all around the globe. Indeed, the papers here collected are highly diverse in terms of the topics addressed and the number of countries where researchers are based. In particular, in the case of nine papers, the laboratories involved were based in eight different countries (Italy, Japan, China, Iran, Iraq, USA, Austria and Switzerland), resulting in a close collaboration to produce high-quality research in the root field.

Simiele et al. [17] investigated the application of organic amendments to improve root traits of poplar cuttings to be used in afforesting or reforesting activities. They found that compost alone seems to be the best solution in both ameliorating substrate characteristics and increasing plant growth, highlighting the great potential for its proper and effective application in large-scale forest restoration strategies. Todo et al. [18] aimed to look at a new methodological approach and demonstrated that point data acquired through 3D laser scanner measurements is a suitable method for fast and accurate reconstruction of root system architecture. Xie et al. [19] found that different *Taxodium* genotypes had different root foraging abilities for phosphorus suggesting that breeding and screening for fine-tuning varieties may help to enhance afforestation success in P-limited areas. He et al. [20] analyzed the concentration of critical secondary metabolites such as organic acids in root tissues of *Taxodium distichum* and *Salix matsudana* in response to cyclical flooding. The authors found that organic acids concentration in the roots of the studied species were mainly influenced by winter flooding. In particular, the exotic species *T. distichum* showed a more stable metabolism of organic acids, while the native species *S. matsudana* responded more actively to long-term winter flooding. Amoli Kondori et al. [21] investigated the effect of different sized forest gaps on fine root dynamics and chemical composition six years

Citation: Montagnoli, A. Adaptation of the Root System to the Environment. *Forests* **2022**, *13*, 595. https://doi.org/10.3390/f13040595

Received: 6 April 2022
Accepted: 7 April 2022
Published: 10 April 2022

after logging. These authors highlighted how, in the medium term and within the adopted size range, the fine root system can recover to pre-harvest conditions in terms of standing biomass and morphological traits. Sugai et al. [22] evaluated the impact of soil compaction and N loading on hybrid larch (*Larix gmelinii* var. japonica × *L. kaempferi*) seedling roots. Outcomes of the study revealed that seedling root development was associated with soil recovery after compaction, and furthermore that under soil compaction N loading promoted root development. Lak et al. [23] in their study addressed the plasticity of fine root traits concerning both interspecific and intraspecific competition of *Acer pseudoplatanus* L. and *Fagus sylvatica* L. seedlings in nutrient-rich soil patches. The authors observed that both con- and allospecific roots had similar effects on target root growth and most trait values. Both species showed highly species- but not competitor-specific root traits plasticity. Deljouei et al. [24] evaluated root tensile force for two temperate tree species within the Caspian Hyrcanian Ecoregion (i.e., *Fagus orientalis* L. and *Carpinus betulus* L.) at three different elevations, for three diameters at breast height (DBH) classes, and at two slope positions. They identified tree species and DBH as the main factors affecting variability in fine root tensile force, with the roots of *F. orientalis* being stronger than those of *C. betulus* in the large DBH class. Finally, Wang et al. [25] carried out a comprehensive analysis of the TIFY gene family expression profiles in *Populus trichocarpa* root tissues under phytohormone treatment and abiotic stresses, such as drought, heat, and cold. The qRT–PCR analysis revealed that almost all PtrTIFY genes responded to at least one abiotic stress or phytohormone treatment, revealing their important role in these functional responses. Together, these papers provide a large perspective on the knowledge advancement in plant root research, but at the same time give a clear indication of the gaps that still need to be closed to improve root related issues, such as methodology, plasticity, gene activation, forest management, and enhancement of afforestation programs.

Funding: This research was funded by University of Insubria.

Conflicts of Interest: The author declares no conflict of interest.

References

1. Montagnoli, A.; Dumroese, K.D.; Terzaghi, M.; Onelli, E.; Scippa, G.S.; Chiatante, D. Seasonality of fine root dynamics and activity of root and shoot vascular T cambium in a *Quercus ilex* L. forest (Italy). *For. Ecol. Manag.* **2019**, *431*, 26–34. [CrossRef]
2. Dumroese, K.D.; Terzaghi, M.; Chiatante, D.; Scippa, G.S.; Lasserre, B.; Montagnoli, A. Functional Traits of Pinus ponderosa Coarse Roots in Response to Slope Conditions. *Front. Plant Sci.* **2019**, *10*, 947. [CrossRef] [PubMed]
3. Montagnoli, A.; Terzaghi, M.; Chiatante, D.; Scippa, G.S.; Lasserre, B.; Dumroese, K.D. Ongoing modification to root system architecture of *Pinus ponderosa* growing on a sloped site revealed by tree-ring analysis. *Dendrochronologia* **2019**, *58*, 125650. [CrossRef]
4. Montagnoli, A.; Lasserre, B.; Sferra, G.; Chiatante, D.; Scippa, G.S.; Terzaghi, M.; Dumroese, K.D. Formation of Annual Ring Eccentricity in Coarse Roots within the Root Cage of Pinus ponderosa Growing on Slopes. *Plants* **2020**, *9*, 181. [CrossRef]
5. Montagnoli, A.; Terzaghi, M.; Scippa, G.S.; Chiatante, D. Heterorhizy can lead to underestimation of fine-root production when using mesh-based techniques. *Acta Oecol.* **2014**, *59*, 84–90. [CrossRef]
6. Baesso, B.; Chiatante, D.; Terzaghi, M.; Zenga, D.; Nieminen, K.; Mahonen, A.P.; Siligato, R.; Helariutta, Y.; Scippa, G.S.; Montagnoli, A. Transcription factors PRE3 and WOX11 are involved in the formation of new lateral roots from secondary growth taproot in *A. thaliana*. *Plant Biol.* **2018**, *20*, 426–432. [CrossRef]
7. Baesso, B.; Terzaghi, M.; Scippa, G.S.; Montagnoli, A. WOX genes expression during the formation of new lateral roots from secondary structures in *Populus nigra* (L.) taproot. *Sci. Rep.* **2020**, *10*, 18890. [CrossRef]
8. De Zio, E.; Montagnoli, A.; Karady, M.; Terzaghi, M.; Sferra, G.; Antoniadi, I.; Scippa, G.S.; Ljung, K.; Chiatante, D.; Trupiano, D. Reaction Wood Anatomical Traits and Hormonal Profiles in Poplar Bent Stem and Root. *Front. Plant Sci.* **2020**, *11*, 59098. [CrossRef]
9. Montagnoli, A.; Baronti, S.; Danieli, A.; Chiatante, D.; Scippa, G.S.; Terzaghi, M. Pioneer and fibrous root seasonal dynamics of Vitis vinifera L. are affected by biochar application to a low fertility soil: A rhizobox approach. *Sci. Total Environ.* **2021**, *751*, 141455. [CrossRef]
10. Montagnoli, A.; Terzaghi, M.; Di Iorio, A.; Scippa, G.S.; Chiatante, D. Fine-root morphological and growth traits in a Turkey-oak stand in relation to seasonal changes in soil moisture in the Southern Apennines, Italy. *Ecol. Res.* **2012**, *27*, 1015–1025. [CrossRef]
11. Montagnoli, A.; Terzaghi, M.; Magatti, G.; Scippa, G.S.; Chiatante, D. Conversion from coppice to high stand increase soil erosion in steep forestland of European beech. *Reforesta* **2016**, *2*, 60–75. [CrossRef]

12. Danjon, F.; Fourcaud, T.; Bert, D. Root architecture and wind-firmness of mature *Pinus pinaster*. *New Phytol.* **2005**, *168*, 387–400. [CrossRef] [PubMed]
13. Montagnoli, A.; Terzaghi, M.; Di Iorio, A.; Scippa, G.S.; Chiatante, D. Fine-root seasonal pattern, production and turnover rate of European beech (*Fagus sylvatica* L.) stands in Italy Prealps: Possible implications of coppice conversion to high forest. *Plant Biosyst.* **2012**, *146*, 1012–1022. [CrossRef]
14. Terzaghi, M.; Montagnoli, A.; Di Iorio, A.; Scippa, G.S.; Chiatante, D. Fine-root carbon and nitrogen concentration of European beech (*Fagus sylvatica* L.) in Italy Prealps: Possible implications of coppice conversion to high forest. *Front. Plant Sci.* **2013**, *4*, 192. [CrossRef] [PubMed]
15. Di Iorio, A.; Montagnoli, A.; Terzaghi, M.; Scippa, G.S.; Chiatante, D. Effect of tree density on root distribution in *Fagus sylvatica* stands: A semi-automatic digitising device approach to trench wall method. *Trees* **2013**, *27*, 1503–1513. [CrossRef]
16. Montagnoli, A.; Chiatante, D.; Godbold, D.L.; Koike, T.; Rewald, B.; Dumroese, R.K. Editorial: Modulation of Growth and Development of Tree Roots in Forest Ecosystems. *Front. Plant Sci.* **2022**, *13*, 850163. [CrossRef]
17. Simiele, M.; De Zio, E.; Montagnoli, A.; Terzaghi, M.; Chiatante, D.; Scippa, G.S.; Trupiano, D. Biochar and/or Compost to Enhance Nursery-Produced Seedling Performance: A Potential Tool for Forest Restoration Programs. *Forests* **2022**, *13*, 550. [CrossRef]
18. Todo, C.; Ikeno, H.; Yamase, K.; Tanikawa, T.; Ohashi, M.; Dannoura, M.; Kimura, T.; Hirano, Y. Reconstruction of Conifer Root Systems Mapped with Point Cloud Data Obtained by 3D Laser Scanning Compared with Manual Measurement. *Forests* **2021**, *12*, 1117. [CrossRef]
19. Xie, R.; Hua, J.; Yin, Y.; Wan, F. Root Foraging Ability for Phosphorus in Different Genotypes Taxodium 'Zhongshanshan' and Their Parents under Phosphorus Deficiency. *Forests* **2021**, *12*, 215. [CrossRef]
20. He, X.; Wang, T.; Wu, K.; Wang, P.; Qi, Y.; Arif, M.; Wei, H. Responses of Swamp Cypress (*Taxodium distichum*) and Chinese Willow (*Salix matsudana*) Roots to Periodic Submergence in Mega-Reservoir: Changes in Organic Acid Concentration. *Forests* **2021**, *12*, 203. [CrossRef]
21. Kondori, A.A.; Vajari, K.A.; Feizian, M.; Montagnoli, A.; Di Iorio, A. Gap Size in Hyrcanian Forest Affects the Lignin and N Concentrations of the Oriental Beech (*Fagus orientalis* Lipsky) Fine Roots but Does Not Change Their Morphological Traits in the Medium Term. *Forests* **2021**, *12*, 137. [CrossRef]
22. Sugai, T.; Yokoyama, S.; Tamai, Y.; Mori, H.; Marchi, E.; Watanabe, T.; Satoh, F.; Koike, T. Evaluating Soil–Root Interaction of Hybrid Larch Seedlings Planted under Soil Compaction and Nitrogen Loading. *Forests* **2020**, *11*, 947. [CrossRef]
23. Lak, Z.A.; Sandén, H.; Mayer, M.; Godbold, D.L.; Rewald, B. Plasticity of Root Traits under Competition for a Nutrient-Rich Patch Depends on Tree Species and Possesses a Large Congruency between Intra- and Interspecific Situations. *Forests* **2020**, *11*, 528. [CrossRef]
24. Deljouei, A.; Abdi, E.; Schwarz, M.; Majnounian, B.; Sohrabi, H.; Dumroese, R.K. Mechanical Characteristics of the Fine Roots of Two Broadleaved Tree Species from the Temperate Caspian Hyrcanian Ecoregion. *Forests* **2020**, *11*, 345. [CrossRef]
25. Wang, H.; Leng, X.; Xu, X.; Li, C. Comprehensive Analysis of the TIFY Gene Family and Its Expression Profiles under Phytohormone Treatment and Abiotic Stresses in Roots of *Populus trichocarpa*. *Forests* **2020**, *11*, 315. [CrossRef]

Article

Biochar and/or Compost to Enhance Nursery-Produced Seedling Performance: A Potential Tool for Forest Restoration Programs

Melissa Simiele [1,†], Elena De Zio [1,†], Antonio Montagnoli [2], Mattia Terzaghi [3], Donato Chiatante [2], Gabriella Stefania Scippa [1] and Dalila Trupiano [1,*]

1 Department of Biosciences and Territory, University of Molise, 86090 Pesche, Italy; melissa.simiele@unimol.it (M.S.); elena.dezio@studenti.unimol.it (E.D.Z.); scippa@unimol.it (G.S.S.)
2 Department of Biotechnology and Life Sciences, University of Insubria, 21100 Varese, Italy; antonio.montagnoli@uninsubria.it (A.M.); donato.chiatante@uninsubria.it (D.C.)
3 Department of Chemistry and Biology "A. Zambelli", University of Salerno, 84084 Fisciano, Italy; mterzaghi@unisa.it
* Correspondence: dalila.trupiano@unimol.it
† These authors contributed equally to this work.

Abstract: Today, the use of nursery-produced seedlings is the most widely adopted method in forest restoration processes. To ensure and enhance the performance of transplanting seedlings into a specific area, soil amendments are often used due to their ability to improve soil physicochemical properties and, in turn, plant growth and development. The aim of the present study was to evaluate *Populus euramericana* growth and development on a growing substrate added with biochar and compost, both alone and in combination. To accomplish this aim, a pot experiment was performed to test biochar and/or compost effects on growing substrate physicochemical characteristics, plant morpho-physiological traits, and plant phenology. The results showed that biochar and/or compost improved growing substrate properties by increasing electrical conductivity, cation exchange capacity, and nutrient concentrations. On the one hand, these ameliorations accelerated poplar growth and development. On the other hand, amendments did not have positive effects on some plant morphological traits, although compost alone increased plant height, and very fine and fine root length. The combined use of biochar and compost did not show any synergistic or cumulative beneficial effects and led to a reduction in plant growth and development. In conclusion, compost alone seems to be the best solution in both ameliorating substrate characteristics and increasing plant growth, highlighting the great potential for its proper and effective application in large-scale forest restoration strategies.

Keywords: fine roots; morphological attributes; physiological analysis; *Populus euramericana*; reforestation

Citation: Simiele, M.; De Zio, E.; Montagnoli, A.; Terzaghi, M.; Chiatante, D.; Scippa, G.S.; Trupiano, D. Biochar and/or Compost to Enhance Nursery-Produced Seedling Performance: A Potential Tool for Forest Restoration Programs. *Forests* **2022**, *13*, 550. https://doi.org/10.3390/f13040550

Academic Editor: Timo Domisch

Received: 25 February 2022
Accepted: 29 March 2022
Published: 31 March 2022

1. Introduction

Forest ecosystems cover thirty-one percent of the global land area [1] and are important for human livelihoods, climate stability, and biodiversity conservation [2]. However, forests are still under constant threat because of deforestation and forest degradation, which continue at an alarming rate [3]. Since 1990, it has been estimated that 420 million hectares of forest have been lost through conversion to other land uses [4]. Between 2015 and 2020, the rate of deforestation was estimated at 10 million hectares per year, and 15 billion trees are cut down every year [5]. Currently, forests are no longer endangered solely by deforestation practices, but in an increasingly warming world, phenomena such as windstorms, heat waves, and droughts pose new and severe threats [6]. Recently, it has been documented that forest death in southwestern Australia has been directly linked to a heat wave that occurs during a "warmer drought" [7].

At the international level, the sustainable management of reforested and afforested sites has been promoted to contribute to maintaining forest status, preserving biodiversity [8], guaranteeing ecosystem services, reducing climate change impact [9], and avoiding soil degradation and desertification [10]. Today, many countries are increasing reforestation and afforestation efforts to remediate the many practices of forest clearance that occurred in the past and forest losses caused by harmful events related to climate change [11].

However, both reforestation and afforestation processes are complex, time-demanding [12], and expensive considering the costs of soil preparation and fertilization, plant tree species selection and acquirement, maintenance, and management practices [13]. Moreover, forest restoration projects have high failure rates due to the high mortality of plants before reaching maturity [14]. For these reasons, it has become increasingly important to develop methods that may increase seedling survival, growth, and vigor while reducing labor costs [15]. In particular, it is important to ensure that planted seedlings have high survival rates and good growth [16], and the use of nursery-produced seedlings is the most widely adopted method in forest restoration processes [17]. Successful plant establishment depends greatly on decisions made prior to planting, which must take into account species and site characteristics [18] and that nursery-produced seedlings will face a variety of stress factors after nursery production [19].

Several methods have been tested in order to ensure and increase the performance of seedling outplanting in deforested and degraded lands, in which salinity, low water holding capacity, and lack of nutrient availability affect the realization of reforestation and afforestation methods [12,20,21]. To overcome these limitations, soil amendments can be used to improve the physicochemical properties of soils [22] and guarantee land restoration success [23].

Among the different amendments, biochar—a charcoal produced by the pyrolysis of organic waste feedstocks, such as manure and crop residues—has received attention [24]. It is a carbon sequestration agent able to reduce atmospheric CO_2 concentrations and is an excellent soil conditioner [25]. Indeed, it is effective in improving degraded sites because biochar application increases soil aggregate stability and water holding capacity by enhancing soil pore characteristics and water retention [26]. In addition to soil responses, biochar may yield a wide range of benefits for plant germination, growth, productivity, and survival, and stress management [20]. Several mechanisms may enhance plant development in response to soil biochar additions, including: (i) reduction of nutrient leaching and improvement of plant-available nutrients; (ii) release of carbon, nitrogen, phosphorus, and potassium; and (iii) increase of soil biota density and diversity [27]. Thus, it is evident that biochar has several properties of particular interest from the perspective of forest restoration. First, its recalcitrance implies that biochar added in the context of a restoration project will not rapidly decompose [28]. Secondly, biochar positively acts on plant growth and survival in highly degraded soils, which are frequent in the context of forest restoration [29]. It is also particularly effective in adsorbing a wide range of materials that are either generally toxic or adverse to plant growth at high concentrations, including metals, salts, polyaromatic hydrocarbons, and residual herbicides [30]. Lastly, biochar may be relatively easily and economically generated from locally available feedstocks, and thus it offers important potential advantages in both economic and logistic terms [15].

Another soil conditioner that may be profitably used in forest restoration is compost, a fertilizer able to improve soil quality by: (i) incorporating organic matter [31,32], nutrients, and electrolytes into soil [33]; and (ii) enhancing soil structure, density, and porosity [34], which increase water retention capacity and reduce soil erosion and nutrient leaching [35]. In turn, such compost properties result in the compost having positive effects on plant growth, even because compost amendment may activate a wide range of natural disease suppressiveness mechanisms against plant pathogens [36]. Moreover, by enhancing the carbon storage capacity in the soil, compost might be used in reducing global warming [37].

Numerous studies have suggested the application of biochar in combination with compost as a promising strategy to promote plant growth and performance, having positive synergistic effects on soil properties [23,38–40]. However, this positive effect is strictly related to specific soil characteristics, plant species, amendment application rates, and feedstock [20]. Thus, more experiments are needed to accurately test synergism using quantifiable metrics to determine what a "target" seedling might be [41].

The Target Plant Concept (TPC) is an effective framework for defining, producing, and managing plant material (e.g., seeds, cuttings, and seedlings) based on characteristics appropriate for a specific site [16]. These characteristics are often scientifically derived from test factors that are linked to outplanting performance, such as seedling morphology and physiology [42], genetic origin, and the ability to overcome limiting conditions on outplanting sites [43]. Commonly measured morphological attributes include bud development, dry weight fraction, stem height and diameter, and root development, which are used in seedling quality assessment programs to monitor plant growth and survival [10]. Seedling morphological attributes cannot be used alone to assess seedling quality because morphology does not describe physiological vigor [44], and, as mentioned before, seedling morphology is combined with the assessment of physiological attributes (e.g., nutrient status, root growth potential, stress tolerance) to relate seedling quality at lifting to field performance after planting [41].

In line with the above, our study examined the potential for biochar and compost amendments, alone or in combination, to have benefits in nursery-produced seedling systems, concomitantly enhancing a range of soil properties, and improving plant growth and development. The main goal of this study was to conduct targeted research on poplar plants to highlight the potential beneficial use of biochar and/or compost in reforested environments. Poplar and its hybrids can be used to create economic benefits and improve environmental quality in forestry and agroforestry worldwide [45,46]. They have shown the capacity for rapid biomass accretion [47] and are currently assuming growing importance for timber and bioenergy production [48]. The adoption of poplar species into the agroforestry system has the added benefit of sequestering carbon emitted from agricultural practices [49]. Thus, the outcomes of this research might enlarge the understanding of soil amendments as ameliorants of soil properties, as well as survival of key species in reforestation and afforestation programs [50].

To accomplish the objective of the present research, biochar and compost, both alone and in combination, were tested on hybrid poplar seedlings with the aim of determining if the addition of biochar and/or compost to a growing media may (i) enhance soil physicochemical properties and (ii) improve plat morphophysiological traits, with particular regard to root morphology and development.

2. Materials and Methods

2.1. Experimental Design and Growth Characteristics

Two-year-old woody cuttings of the hybrid poplar clone I-214, *Populus deltoides* × *Populus nigra* (*Populus euramericana* (Dode) Guinier) were rooted in 3 L pots containing vermiculite and kept in a growth chamber under controlled conditions (25 °C air temperature, 50–70% relative humidity, 15-h photoperiod) for 60 days.

Then, 40 homogeneously rooted cuttings, with similar morphological traits (e.g., size), were selected for the experiment and transplanted, separately, into 20 L plastic pots. These pots were filled with four different growing medium combinations (hereafter also called growing substrates or treatments), and ten replicates were set up for each of them. *P. euramericana* plants were grown for each treatment in a greenhouse for 12 months. The experimentation began in July and ended in July of the following year. In the 12 months of experimentation, hereinafter, are indicated by the acronyms ranging from T0 to T12 (Figure 1). Growth occurred under a controlled water regime and natural photoperiod and temperature (for minimum, maximum, and average temperature values see Table S1 in Supplementary Material), and the plants were arranged in a randomized complete block design and rotated to a different position within the block throughout the trial. The pots were fully irrigated to prevent water stress (twice a day, as required), and a suspended net was used to reduce exposure to sunlight.

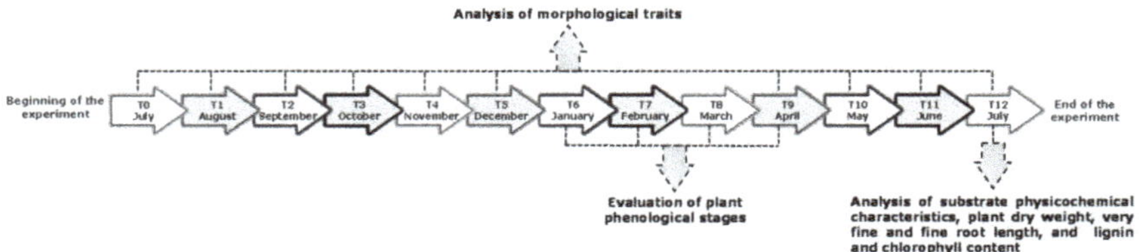

Figure 1. Experimental schedule with the main analyses and the sampling times (months).

The four treatments were: (i) control mix (Ctrl), composed of soil and sand (1:1, *v:v*); (ii) biochar mix (B), composed of soil and sand (1:1) plus biochar; (iii) compost mix (Co), composed of soil and sand (1:1) plus compost; and (iv) biochar and compost mix (BCo), composed of soil and sand (1:1) plus biochar and compost. For treated pots, biochar and compost were added to the soil and sand mixture at a concentration of 25 g·kg^{-1} of dry substrate (application rate of 2.5% *w*/*w*).

2.2. Soil, Biochar, and Compost Characteristics

The soil was collected from an uncultivated pasture area, located in Pesche (Molise, Italy), with a floral composition predominantly of graminoid grasses, not under a rotation system, and that includes hedges. This area is mainly used for grazing, but the fodder is harvested mechanically. The soil was loamy mixed mesic soil, according to the United States Department of Agriculture (USDA) classification [51]. As reported previously, the soil was mixed with commercial sand, and this mixture was characterized to determine the pre-planting physicochemical properties. The soil–sand mix was moderately subalkaline with a silt loam texture according to the USDA classification. Moreover, it was characterized by low electrical conductivity (EC), cation exchange capacity (CEC), and nitrogen and carbon content. For the experiment, the mixture of the soil and sand was air-dried for 72 h, weighed, finely crushed, and then mixed thoroughly before packing lightly in the pots on top of the pebbles placed on the base to improve drainage.

The biochar used was a commercial charcoal (provided by Romagna Carbone s.n.c., Bagnacavallo, Italy), obtained from orchard pruning biomass through a slow pyrolysis process with an average residence time of 3 h, at a temperature of 500 °C, in a kiln of 2.2 m in diameter, and holding around 2 ton of feedstock.

The compost was a commercial product (composted olive mill residues) prepared in a standardized experimental composting process reported by Alfano et al. [52]. Briefly, compost was prepared by mixing humid olive husks from a two-phase extraction plant with olive leaves (8% *w*/*w*); one-year-old, humid, and composted husks (25% *w*/*w*) were then added to this mixture. A complete overview of both biochar and compost characteristics is reported by Trupiano et al. [53].

2.3. Growing Substrate Analysis

At the end of the experiment (T12), after 12 months of plant growth, soil samples were collected (Figure 1) and air dried at 20–25 °C for 72 h. The moisture content was calculated according to the Black method [54] as the difference in sample weight before and after oven drying to a constant weight at 105 °C. The pH was measured in H_2O and 0.01 M $CaCl_2$ using a pH meter (Eutech Instruments, Thermo Fisher Scientific, Waltham, MA, USA) and a 1:2.5 soil weight:extract-volume ratio. The alkalinity of samples with a pH value greater than 7.0 was determined by titrimetry according to Rayment and Higginson [55]. Electrical conductivity (EC) was determined by a conductivity meter (Cond 510, XS Instruments, Carpi, Italy) on a 1:5 soil:water suspension [56]. Cation exchange capacity (CEC) was assessed according to the method of Mehlich [57] using $BaCl_2$. Total organic carbon (TOC) and total nitrogen (N_{tot}) contents were determined by dry combustion [58] using a CHN elemental analyzer (Mod 1500, series 2, Carlo Erba Instruments, Cornaredo, Italy). In the case of TOC, combustion was carried out after the complete removal of inorganic carbon with acid. Total phosphorus (P_{tot}) was detected by spectrophotometry (UV-1601 Shimadzu) according to the test method described by Bowman [59], while available phosphorus (P_{av}) was extracted by a $NaHCO_3$ solution at pH 8.5 and evaluated by spectrophotometry according to the Olsen test method [60]. Particle size distribution (also named soil texture) was quantified by hydrometer analysis through a modification of the Bouyoucos method [61] (according to Beretta et al. [62]) on samples previously dry-sieved at 2 mm. The fraction with a diameter smaller than 2 mm was treated with H_2O_2 and wet sieved at 200 μm, 50 μm, and 20 μm. Measurements of density were carried out by a hydrometer on samples smaller than 20 μm previously dispersed with sodium hexametaphosphate solution. Moreover, in order to quantify the large fraction of macro aggregates, particles that did not pass through the 2 mm sieve were treated with sodium hexametaphosphate solution to disrupt aggregates, and subsequently, the difference in weight before and after wet sieving at 2 mm was measured [63].

2.4. Plant Analysis

2.4.1. Morphological Traits

From the beginning (T0) to the end of the experiment (T12), plant growth was monitored monthly by measuring morphological traits. More detailed, morphological traits were assessed at time points T1–T5, corresponding to the months of August–December, and at time points T9–T12, corresponding to the months of April–July. The T1–T5 time points were equivalent to the months preceding the dormancy phase (January—T6, February—T7, and March—T8) and T9–T12 to the months of the vegetative phase of *P. euramericana* (Figure 1). For morphological traits, the variations in plant height and leaf number (Δ) were determined. Additionally, the main leaf parameters were assessed: leaf area, perimeter, length, and width. Image J 1.8.0 software (Wayne Rasbanb National Institute of Health, Bethesda, MD, USA; http://rsb.info.nih.gov/ij/ (accessed on 20 November 2020)) was used for these analyses. The aboveground (leaves and stems) and belowground (roots) dry weight, at time T12, were also determined after two days of drying in an oven at 80 °C, and stem/root ratio (S/R ratio) was calculated in terms of dry weight. Furthermore, roots were analyzed using WinRhizo Pro V. 2007d (Regent Instruments Inc. Quebec, QC, Canada). In particular, roots were divided into two groups on the basis of their diameter to distinguish very fine roots (diameter measurement between 0 and 0.5 mm) and fine roots (diameter measurement between 0.5 and 2 mm) [21,64].

2.4.2. Phenology

Phenological stage assessment was performed on poplar plants evaluating apical bud development in the three months of dormancy (January—T6, February—T7, and March—T8) and in the first month of active vegetative growth (April—T9) (Figure 1). Bud development was quantified using six levels of morphology scores (0–6) according to Trupiano et al. [65]. Briefly, a minimal score (0) was given to the winter bud and a maximum score (6) to the flushing buds, with a growing stem.

2.4.3. Lignin and Chlorophyll Content

At the end of the experiment (T12), lignin and chlorophyll contents were also assessed (Figure 1).

Lignin content was measured using the protocol of Doster and Bostock [66] with a few modifications, as detailed in Trupiano et al. [67]. Briefly, lignin amount within each sample was calculated by measuring the absorbance at 280 nm, using a specific absorbance coefficient of 6.0 $L \cdot g^{-1}$ cm. Since this specific absorbance coefficient provides only an approximate conversion [66], the samples with the highest lignin content value were used as a standard in the relative measurements of lignin concentration of the other samples.

Chlorophyll content was measured in three randomly sampled leaf discs (10 mm in diameter). Extraction was performed with N, N dimethylformamide (DMF) for 48 h at 4 °C, in the dark at a ratio of 1:20 (plant material:solvent, *w:v*). The extinction coefficients proposed by Inskeep and Bloom [68] were used for quantification by spectrophotometric analysis. The following equations were used: Chl a = $12.70_{A664 \cdot 5} - 2.79_{A647}$; Chl b = $20.70_{A647} - 4.62_{A664 \cdot 5}$; total Chl = $17.90_{A647} + 8.08_{A664 \cdot 5}$ where A is absorbance in 1.00 cm cuvettes.

2.5. Statistical Analysis

To evaluate the effect of each treatment on the conducted tests, all statistical analyses were performed with the R software version 3.4.3 (R Development Core Team 2017). The normality and homoscedasticity of the data were assessed with the Shapiro and Bartlett tests, respectively. The mean values were compared using the parametric analysis of variance test (ANOVA) for normal data or the non-parametric Kruskal test for non-normal data. Following this, a post hoc test (Tukey HSD or pairwise Wilcox tests, respectively) was performed.

3. Results

3.1. Growing Substrate Characteristics

At the end of the experiment (T12), the moisture content of the medium with the compost addition (Co substrate) (1.15 $g \cdot kg^{-1}$) was 32% lower compared to the control (1.52 $g \cdot kg^{-1}$). However, no significant difference was present among the four different substrates in pH and alkalinity values (Table 1).

Table 1. Main physicochemical characteristics determined after 12 months of *P. euramericana* grown on four different substrates (Ctrl, B, Co, and BCo). Different letters indicate significant difference ($p < 0.05$) ($n = 3 \pm SE$). Ctrl: control mix of soil and sand; B: mix of soil and sand + biochar; Co: mix of soil and sand + compost; BCo: mix of soil and sand + biochar + compost.

	Ctrl	B	Co	BCo
Moisture content ($g \cdot kg^{-1}$)	1.52 ± 0.06 a	1.43 ± 0.05 ab	1.15 ± 0.01 b	1.57 ± 0.14 a
pH	7.4 ± 0.0 a	7.4 ± 0.0 a	7.4 ± 0.0 a	7.4 ± 0.0 a
Alkalinity (% $CaCO_3$)	10.00 ± 0.29 a	10.57 ± 0.49 a	9.53 ± 0.33 a	10.83 ± 0.24 a
EC ($dS \cdot m^{-1}$)	0.233 ± 0.005 c	0.247 ± 0.006 bc	0.447 ± 0.002 a	0.340 ± 0.025 b
CEC (cmol(+)$\cdot kg^{-1}$)	6.49 ± 0.36 b	8.62 ± 0.27 a	9.69 ± 0.18 a	9.60 ± 0.23 a
TOC ($g \cdot kg^{-1}$)	12.26 ± 0.94 b	13.38 ± 1.27 b	23.41 ± 1.28 a	23.54 ± 0.69 a
N_{tot} ($g \cdot kg^{-1}$)	0.23 ± 0.01 d	0.52 ± 0.01 c	1.60 ± 0.05 b	2.06 ± 0.01 a
P_{tot} ($mg \cdot kg^{-1}$)	129.33 ± 4.52 c	215.00 ± 3.24 b	301.00 ± 2.94 a	324.00 ± 5.93 a
P_{av} ($mg \cdot kg^{-1}$)	<12 b	<12 b	16.90 ± 1.49 a	18.63 ± 1.01 a
Particle size distribution *:				
ø < 2 μm (%) (clay)	8.90 ± 0.36 b	6.27 ± 0.18 c	13.50 ± 0.18 a	9.63 ± 0.96 b
2 < ø < 20 μm (%) (silt)	53.80 ± 1.61 a	52.70 ± 1.19 a	53.17 ± 1.72 a	54.63 ± 1.70 a
20 < ø < 50 μm (%) (very fine sand)	2.63 ± 0.91 a	1.93 ± 0.05 a	1.90 ± 0.16 a	1.60 ± 0.29 a
50 < ø < 200 μm (%) (fine sand)	8.93 ± 0.55 ab	9.50 ± 0.35 a	7.90 ± 1.38 ab	6.73 ± 0.33 b
200 μm < ø < 2 mm (%) (coarse sand)	6.83 ± 0.49 a	7.20 ± 0.41 a	8.67 ± 0.23 a	9.10 ± 0.79 a
ø > 2 mm (%) (gravel)	27.80 ± 0.89 a	28.67 ± 0.55 a	28.33 ± 0.58 a	28.00 ± 0.36 a

EC = electrical conductivity ($dS \cdot m^{-1}$); CEC = cation exchange capacity (cmol(+)$\cdot kg^{-1}$); TOC = total organic carbon ($g \cdot kg^{-1}$); N_{tot} = total nitrogen content ($g \cdot kg^{-1}$); P_{tot} = total phosphorus content ($mg \cdot kg^{-1}$); P_{av} = available phosphorous content ($mg \cdot kg^{-1}$). * The classes of this particle size distribution were defined by the International Society of Soil Science (ISSS) system [69].

Electrical conductivity (EC) was increased (2-fold) in the Co treatment (0.447 $dS \cdot m^{-1}$) compared with the Ctrl growth medium (0.233 $dS \cdot m^{-1}$). Cation exchange capacity (CEC) was increased in all three treatments with biochar and/or compost addition (B, Co, and BCo) compared to Ctrl (6.49 cmol(+)$\cdot kg^{-1}$). In detail, CEC was increased by 33%, 49%, and 48% in the B, Co, and BCo treatments compared to the Ctrl medium (Table 1). The contents of total organic carbon (TOC), total nitrogen (N_{tot}), and total (P_{tot}) and available phosphorus (P_{av}) varied significantly and differently among the four treatments. However, the values of these nutrients were all higher in the Co and BCo treatments. Specifically, TOC was 2-fold higher in both the Co and BCo substrates compared to both other two growing media (Ctrl and B). With regard to N_{tot}, it was increased 2-fold in the B substrate (0.52 $g \cdot kg^{-1}$), 3-fold in the Co medium (1.60 $g \cdot kg^{-1}$), and 9-fold in the BCo mix (2.06 $g \cdot kg^{-1}$) with respect to the control (0.23 $g \cdot kg^{-1}$). P_{tot} was also raised by 66% in the B treatment (215 $mg \cdot kg^{-1}$) and was increased 2-fold in the Co (301 $mg \cdot kg^{-1}$) and BCo treatments (324 $mg \cdot kg^{-1}$)

compared with the control (129.33 mg·kg^{-1}). The highest values of P_{av} were measured in the two Co (16.90 mg·kg^{-1}) and BCo treatments (18.63 mg·kg^{-1}) compared to the control growth medium and the B treatment in which there was an amount of phosphorus less than 12 mg·kg^{-1}. Particle size analysis showed that there were differences among the treatments only for the clay and fine sand fractions (Table 1). Soil with the compost (Co substrate) had the highest clay content (13.5%), and in detail, the clay fraction was 52% higher than the control (8.9%). In contrast, the biochar decreased the clay component of the B substrate (6.27%) by 30% compared to the Ctrl. Regarding the fine sand content, the analyses showed only a difference between the sand fraction found in the B treatment (9.5%) and that measured in the BCo mix (6.73%).

3.2. Plant Characteristics

3.2.1. Morphological Traits

After one month of poplar plant growth (T1), the combined addition of the biochar and compost to the growth substrate (BCo) appeared to negatively influence the increase in stem height of *Populus euramericana* seedlings (Figure 2). In the control growing medium (Ctrl), the stem height increased (50%) with respect to the BCo substrate in which the percentage of stem height increased only 27% (Figure 2). At time T2, despite the plant grown on the B, Co, and BCo treatments increasing in stem height of 31%, 7%, and 17%, respectively, the highest increase for this parameter was observed in the Ctrl plants (67%) (Figure 2). From October to March (T3 to T8), as the winter dormancy period advanced, no changes in stem height measurements were recorded in the four different growing media. At times T9 (April) and T10 (May), the development and growth phase restart of *P. euramericana* plants resulted in a significantly higher increases in stem height in both treatments with the compost (Co) and biochar (B) added alone compared to the Ctrl and BCo substrates. Indeed, at time T9, the percentage increase in stem height was 9% in the Co substrate and 5% in the B treatment, while it was 1% in Ctrl and 2% in the BCo mix (Figure 2). At time T10, the increases in stem height were 7% in Co and 6% in B, while stem height increased by 5% and 2% in the Ctrl and BCo substrates, respectively (Figure 2). At time T11, the addition of Co to the soil resulted in the highest percentage of stem growth (59%) compared to the Ctrl (12%), B (14%) and the BCo mix (21%) (Figure 2).

The leaf number changed in poplar seedlings throughout the experiment, and for the four different treatments, it showed highly variable and different trends in each month (Figure 3). The greatest changes between the different treatments of the experiment were observed at T9 and T10. At time T9, the highest increase in leaf number was measured in plants grown on B substrate (559%), followed by 261% and 309% percentage increases observed in the Ctrl and Co growth media, respectively; in the BCo mix, conversely, a strong decrease in leaf number was reported (−92%) (Figure 3). At time T10, the situation changed: the combined addition of biochar and compost (BCo) resulted in the highest percentage of leaf number increase (611%) compared to the Ctrl, B, and Co substrates, in which the percentages were 43%, 23%, and 8%, respectively.

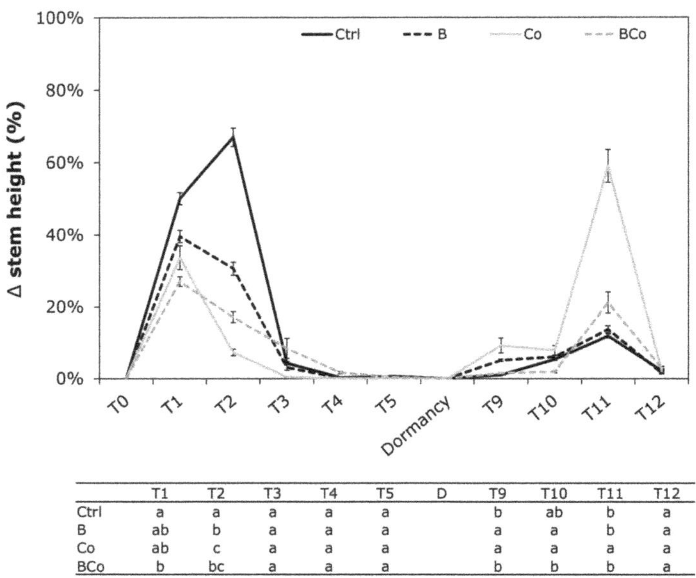

	T1	T2	T3	T4	T5	D	T9	T10	T11	T12
Ctrl	a	a	a	a	a		b	ab	b	a
B	ab	b	a	a	a		a	a	b	a
Co	ab	c	a	a	a		a	a	a	a
BCo	b	bc	a	a	a		b	b	b	a

Figure 2. Variation (Δ) in stem height (%) determined monthly during the 12 months of *P. euramericana* growth on four different substrates (Ctrl, B, Co, and BCo). Different letters indicate significant differences among the treatments for each sampling time ($p < 0.05$) ($n = 10 \pm$ SE). Ctrl: control mix of soil and sand; B: mix of soil and sand + biochar; Co: mix of soil and sand + compost; BCo: mix of soil and sand + biochar + compost. T0 to T12: Sampling times (months).

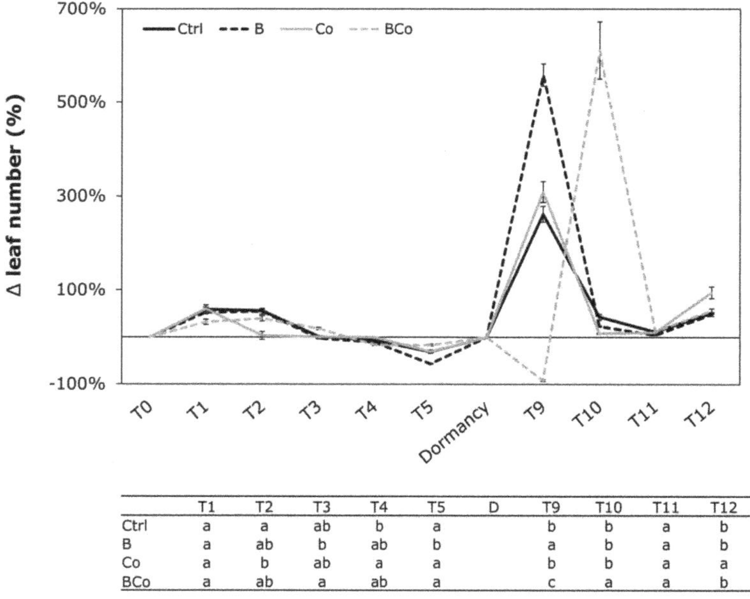

	T1	T2	T3	T4	T5	D	T9	T10	T11	T12
Ctrl	a	a	ab	b	a		b	b	a	b
B	a	ab	b	ab	b		a	b	a	b
Co	a	b	ab	a	a		b	b	a	a
BCo	a	ab	a	ab	a		c	a	a	b

Figure 3. Variation (Δ) in leaf number (%) determined monthly during the 12 months of *P. euramericana* growth on four different substrates (Ctrl, B, Co, and BCo). Different letters indicate significant differences among the treatments for each sampling time ($p < 0.05$) ($n = 10 \pm$ SE). Ctrl: control mix of soil and sand; B: mix of soil and sand + biochar; Co: mix of soil and sand + compost; BCo: mix of soil and sand + biochar + compost. T0 to T12: Sampling times (months).

The main leaf parameters (leaf area, perimeter, length, and width) reported significant differences (Figure 4) in the month immediately preceding the winter dormancy phase (December—T5) and in the two months immediately following dormancy and in which there was active vegetative growth of *P. euramericana* plants (April—T9 and May—T10). In detail, at time T5, all four leaf parameters were lower in B, C, and BCo than in the Ctrl. After the dormancy phase (times T9 and T10), leaf parameters were unchanged in B and Co, whereas they were lower in BCo compared to the Ctrl (Figure 4).

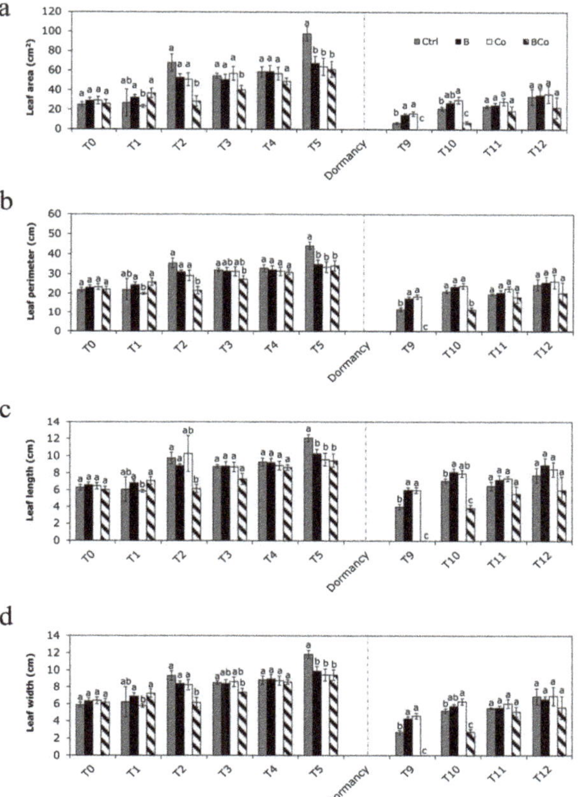

Figure 4. Leaf area (**a**), perimeter (**b**), length (**c**), and width (**d**) determined monthly during the 12 months of *P. euramericana* growth on four different substrates (Ctrl, B, Co, and BCo). Different letters indicate significant differences among the treatments for each sampling time ($p < 0.05$) ($n = 10 \pm$ SE). Ctrl: control mix of soil and sand; B: mix of soil and sand + biochar; Co: mix of soil and sand + compost; BCo: mix of soil and sand + biochar + compost. T0 to T12: Sampling times (months).

Indeed, more specifically, at time T5, the leaf area in the Ctrl was 98 cm^2, while it was almost twice as small in the B (68 cm^2), Co (64 cm^2), and BCo substrate (62 cm^2) (Figure 4a). At time T9, as also shown in Figure 3, poplar plants grown in the BCo treatment were without leaves, and therefore, leaf area was assigned a value of zero (Figure 4a). Conversely, in the B and Co substrates, they were 15 cm^2 and 16 cm^2, respectively, and thus 3 times higher than that measured for the leaves of the Ctrl plants (6 cm^2) (Figure 4a). At time T10, plants grown in the Ctrl, B, and Co growing media had the highest value of leaf area (30 cm^2), which was 7 times bigger than the leaf area of plants grown in the BCo mix (6 cm^2) (Figure 4a).

Regarding the leaf perimeter (Figure 4b), at time T5, the lowest values were found in the B, Co, and BCo treatments (approximately 34 cm compared to the 44 cm measured in the Ctrl plants). At time T9, it was one and a half times bigger in the B (17 cm) and Co (18 cm) treatments with respect to the Ctrl (11 cm), whereas, as mentioned above, plants grown in the BCo mix had no leaves. At time T10, plants grown on the Ctrl, B, and Co substrate had leaf perimeter values of 21 cm, 23 cm, and 24 cm, respectively, which were twice as high as that of plants grown on the BCo mix (11 cm).

The leaf length value was also lower at time T5 for B (10 cm), Co (10 cm), and BCo (9 cm) plants compared to the Ctrl poplar seedlings (12 cm) (Figure 4c). At time T9, it was two-fold higher in B and Co (6 cm in both treatments) compared to the Ctrl and BCo (Figure 4c), whereas at time T10, it was twice as low only in plants grown on the BCo mix (4 cm).

The leaf width (Figure 4d), at time T5, was slightly lower in the B (10 cm), Co (9 cm), and BCo (9 cm) plants than in the Ctrl (12 cm), whereas at time T9, it was approximately twice as large in the B (4 cm) and Co (5 cm) plants compared to the Ctrl plants (Figure 4d). This parameter resulted in the lowest BCo at time T9 and T10 (Figure 4d).

The addition of biochar and compost alone (B and Co treatments) had no effect on the leaf dry weight (Figure 5a). In contrast, the BCo mix decreased the leaf dry weight of *P. euramericana* plants by 40% (9.2 g) compared to the Ctrl growth medium (15.2 g) (Figure 5a). The three treatments—B, Co, and BCo substrates—decreased the stem dry weight by 13%, 12%, and 10%, respectively, with respect to the dry weight measured in the Ctrl (87.8 g) (Figure 5a). However, for the root dry weight, the combination of the two amendments (BCo) resulted in a decrease in dry weight of 47% (11.3 g) compared to the Ctrl (21.5 g) and 45% with respect to the Co treatment (20.4 g) (Figure 5a). The highest stem/root ratio (S/R ratio) was found for plants grown in the BCo mix; these plants had a S/R ratio of 7.42 that was almost two times higher than that found in the Ctrl (4.12) (Figure 5b). For plants grown in the B and Co substrates, the S/R ratio was not significantly different from the Ctrl, although it showed an increased trend in the case of B (5.86) (Figure 5b).

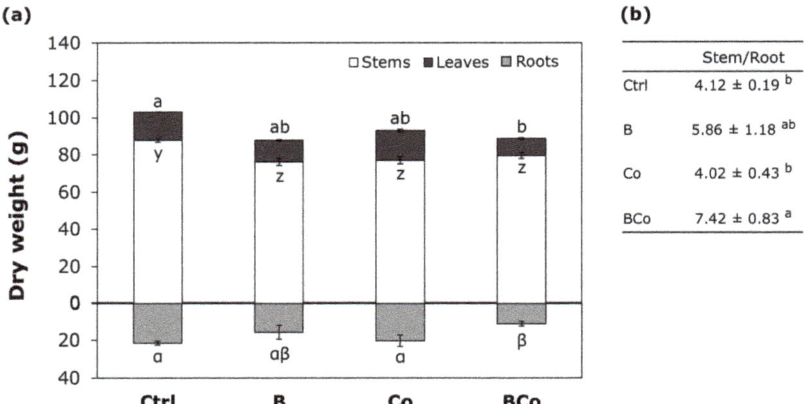

	Stem/Root
Ctrl	4.12 ± 0.19 [b]
B	5.86 ± 1.18 [ab]
Co	4.02 ± 0.43 [b]
BCo	7.42 ± 0.83 [a]

Figure 5. (**a**) Dry weight (g) of the different organs (black box, leaves; white box, stems; gray box, roots) of *P. euramericana* grown on four different substrates (Ctrl, B, Co, and BCo). Different letters indicate a significant difference among substrates ($p < 0.05$) ($n = 3 \pm$ SE): letters a, b for leaves; letters y, z for stems; letters α, β for roots. (**b**) Stem/root ratio calculated as dry weight for stems/dry weight for roots. Different letters indicate a significant difference among substrates ($p < 0.05$) ($n = 3 \pm$ SE). Ctrl: control mix of soil and sand; B: mix of soil and sand + biochar; Co: mix of soil and sand + compost; BCo: mix of soil and sand + biochar + compost.

The addition of compost to the soil (Co treatment) resulted in the highest increase in the length of both very fine and fine roots (Figure 6). More specifically, the Co treatment led to a 33% and 16% increase in very fine root length (12,160 cm) and fine root length (2932 cm) compared to the control growing medium (9115 cm and 2517 cm, respectively; Figure 6). On the other hand, there was a 34% decrease in the BCo plants with respect to the Ctrl substrate (9115 cm) in the case of a very fine root length (Figure 6). Moreover, the two biochar treatments (B and BCo substrates) resulted in a decrease in fine root length of 27% (1837 cm) and 34% (1660 cm) with respect to the Ctrl (2517 cm), respectively (Figure 6).

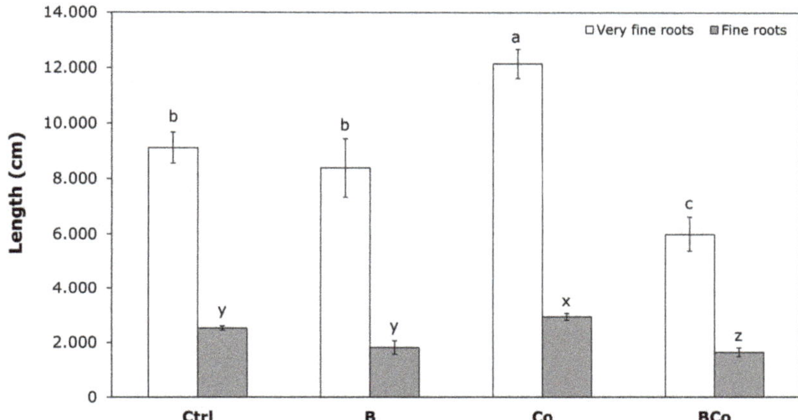

Figure 6. Length (cm) of very fine roots (0 < d < 0.5 mm) and fine roots (0.5 < d < 2 mm) (white box, very fine roots; gray box, fine roots) determined after 12 months of *P. euramericana* growth on four different substrates (Ctrl, B, Co, and BCo). Different letters indicate a significant difference among substrates ($p < 0.05$) ($n = 3 \pm$ SE). Letters a, b, c regard very fine roots; letters x, y, z regard fine roots. Ctrl: control mix of soil and sand; B: mix of soil and sand + biochar; Co: mix of soil and sand + compost; BCo: mix of soil and sand + biochar + compost.

3.2.2. Phenology

The evaluation of *P. euramericana* plant phenological stages (Figure 7) showed that, during winter dormancy, at time T6 (January), all seedlings (100%) in all substrates (Ctrl, B, Co, and BCo) had buds with a morphological score of 0. Successively, at time T7 (February), only seedlings grown in the Ctrl growing medium had buds with a morphological score of 0 (100%). The biochar use (B growth medium) resulted in 20% of the poplar plants having buds with a morphological score of 1. In the two treatments with the compost addition (Co and BCo), the seedlings had buds not only in stage 1 but also in stage 2. In detail, 30% and 10% of plants grown in the Co treatment had buds in stages 1 and 2, respectively. In the BCo mix, 20% and 10% of *P. euramericana* seedlings had buds that were given morphological scores of 1 and 2, respectively.

At time T8 (release from dormancy condition), almost all plants in the control substrate showed buds at stage 1 with only 10% of *P. euramericana* plants at stage 2. Whereas both treatments with the addition of either biochar alone (B substrate) or compost alone (Co substrate) resulted in 20% of plants having stage 1 and those of the remaining 80% having a morphological score of 2. Instead, the combined biochar and compost addition to the soil (BCo mix) resulted in the poplar plants still having stage 0 buds, at time T8. In detail, 20% of the seedlings had buds with the minimum score (0), 70% of *P. euramericana* plants had buds at stage 1, and the remaining 10% showed buds with a morphological score of 2.

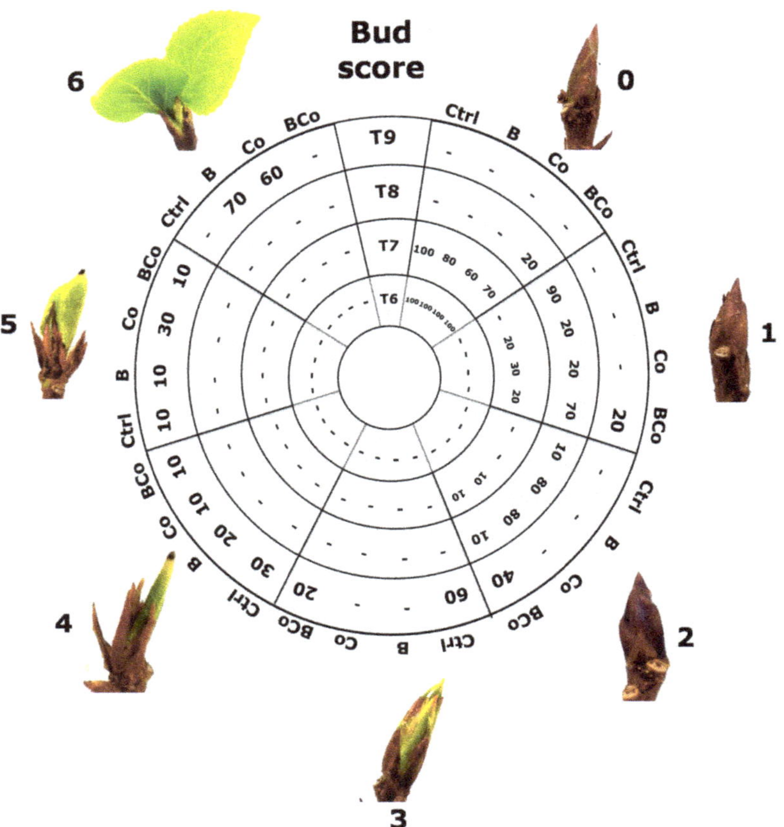

Figure 7. Evaluation of phenological stages of *P. euramericana* grown on four different substrates (Ctrl, B, Co, and BCo) by bud development analysis at different time points: winter dormancy condition (T6, January and T7, February), release from dormancy (T8, March), and active vegetative phase (T9, April). For each time point, the percentage of plants characterized by a specific stage of bud development is reported. Bud development was quantified using six levels of morphology score (0–6); minimal score (0) was given to the winter bud, and maximum score (6) to buds with a growing stem. Ctrl: control mix of soil and sand; B: mix of soil and sand + biochar; Co: mix of soil and sand + compost; BCo: mix of soil and sand + biochar + compost.

During the active vegetative phase (time T9, April), bud development differed among the four distinct treatments. Indeed, in the control growing medium (Ctrl), 60% of the plants had buds at stage 3, and 30% and 10% of the poplar seedlings had buds with morphological scores of 4 and 5, respectively. At time T9, the addition of the biochar alone (B substrate) or compost alone (Co substrate) to the growing substrate resulted in buds characterized by development stages to which scores of 4, 5, and 6 were assigned. In detail, 20% of *P. euramericana* grown in the B substrate had buds with a score of 4, 10% showed buds with a score of 5, and the majority of plants (70%) had buds that were given the maximum score (6). For the Co treatment, 10% of the seedlings were characterized by buds in stage 4, 30% of poplar plants had buds in stage 5, and 60% showed stage 6 buds. Plants grown in the BCo mix had buds with morphological scores ranging from 1 to 5. Specifically, 20% of the plants contained buds with a score of 1, 40% were characterized by buds in stage 2, 20% showed buds with a morphological score of 3, and the remaining plants were half (10%) characterized by buds in stage 4 and half (10%) by buds in stage 5.

3.2.3. Lignin and Chlorophyll Content

The two treatments with the biochar, both added alone (B substrate) and in combination with the compost (BCo substrate), showed the highest lignin content values in the *P. euramericana* roots (Figure 8). In the B treatment, the lignin content was 100%, while in the BCo mix, it was 87%. Thus, in the two biochar treatments, the lignin content was higher than the value found for plants grown in the Ctrl growing medium (70%) and 2 times higher with respect to the content measured for plants developed in the Co substrate (41%).

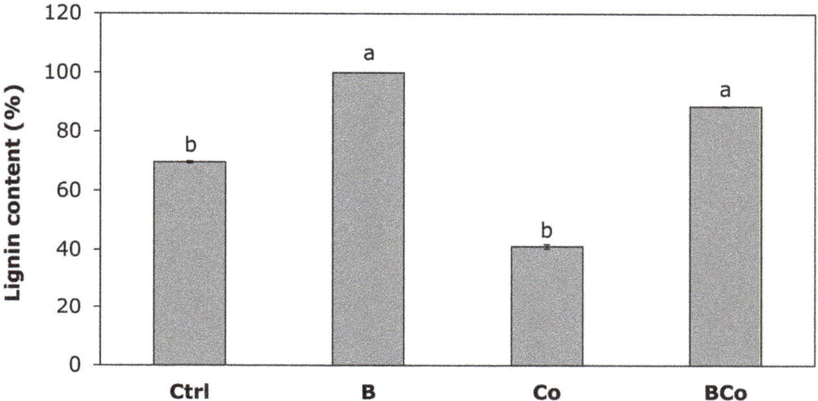

Figure 8. Lignin content (%) determined after 12 months of *P. euramericana* growth on four different substrates (Ctrl, B, Co, and BCo). Lignin content is expressed as a percentage of the value measured in the roots of plants grown in the B substrate (considered 100%). Different letters indicate significant difference ($p < 0.05$) ($n = 3 \pm$ SE). Ctrl: control mix of soil and sand; B: mix of soil and sand + biochar; Co: mix of soil and sand + compost; BCo: mix of soil and sand + biochar + compost.

Regarding the chlorophyll content, Table 2 shows that the values of the total chlorophyll content (Chl), chlorophyll a content (Chl *a*), and chlorophyll b content (Chl *b*) were higher in the B and BCo substrates than in the Ctrl and Co substrates. In detail, the total chlorophyll content in both B (11.51 mg·cm^{-2}) and BCo (11.58 mg·cm^{-2}) was 46% higher than that measured for plants grown in the Ctrl growing medium (7.91 mg·cm^{-2}) and two times higher than the total content measured in the Co treatment (5.47 mg·cm^{-2}). The same trend, as mentioned above, was also observed for Chl *a* and Chl *b* content; however, there were no significant differences between the four different treatments in the ratio of Chl *a* to Chl *b* content.

Table 2. Chlorophyll content (mg·cm^{-2}) of *P. euramericana* grown on four different substrates (Ctrl, B, Co, and BCo). Different letters indicate significant difference ($p < 0.05$) ($n = 10 \pm$ SE).

	Ctrl *	B	Co	BCo
Chl	7.91 ± 0.13 b	11.51 ± 1.99 a	5.47 ± 0.40 b	11.58 ± 1.70 a
Chl *a*	6.17 ± 0.40 b	9.02 ± 0.94 a	4.37 ± 0.37 b	9.04 ± 1.15 a
Chl *b*	1.74 ± 0.27 b	2.48 ± 0.34 a	1.10 ± 0.03 b	2.54 ± 0.55 a
Chl *a*/Chl *b*	3.61 ± 0.79 a	3.63 ± 1.99 a	3.93 ± 0.22 a	3.60 ± 0.33 a

* Ctrl: control mix of soil and sand; B: mix of soil and sand + biochar; Co: mix of soil and sand + compost; BCo: mix of soil and sand + biochar + compost.

4. Discussion

In forest restoration processes, soil amendments (e.g., biochar and compost) may help to ensure and enhance the performance of nursery-produced seedling transplants, increasing plant survival rates and growth [16]. This soil conditioner ability is not only associated with amendment characteristics (e.g., application rate and feedstock), but it is also strictly related and influenced by soil physicochemical properties and the choice of plant species [20].

Several studies have established that pyrolysis temperature and type (fast or slow), together with feedstock choice, influence final biochar physicochemical characteristics and consequently its properties as a soil conditioner [70–73]. Pyrolysis temperature affects biochar longevity, with pyrolysis temperatures higher than 500 °C generally leading to longer-term half-lives [70]. Greater pyrolysis temperatures also led to biochar containing greater carbon concentrations and a specific surface area [71]. However, it appears that feedstock selection has the largest influence on biochar properties [72]. Wood-based biochars have the greatest specific surface area, while crop- and other grass-based biochars appear to have cation exchange capacities greater than other biochars, which could potentially lead to longer-term changes in soil nutrient retention [73].

The selection and combination of feedstock material is also important for the quality of compost, which is produced primarily from animal and agricultural waste. In general, animal wastes, such as cow dung, poultry litter, pig manure, and chicken manure, contain low levels of carbon but high levels of N [74]. Most agricultural wastes, such as crop residues, sawdust, and rice straw, contain large amounts of carbon but low amounts of nitrogen and cause a slower decomposition process [75]. It is important to combine the right materials to achieve compost characterized by correct proportions of C and N. The ideal C/N ratio for compost should be kept at 25–50 because if it is lower, ammonia is emitted, and biological activity can be hindered. On the other hand, at higher ratios, N can be a limiting nutrient, and the composting process will be slow, and too much nitrogen will cause the material to become acidic and smelly [76].

Thus, more experiments are needed to accurately test the effect (negative, positive, synergistic/antagonistic) of soil amendments on selected forestry seedlings, such as poplar plants. In our study, the potential use of biochar and compost, both alone and in combination, on nursery-produced seedlings was evaluated, taking into account changes in growing substrate properties and *Populus euramericana* morphophysiological traits and phenology.

The biochar used was a commercial charcoal obtained from orchard pruning biomass through a slow pyrolysis process at a temperature of 500 °C. This biochar was characterized by an alkaline pH and a high C/N ratio (125.5). The compost used was prepared from olive waste and had a pH close to neutrality (7.5) and an appropriate C/N ratio (28.1).

The study showed that the compost amendment alone produced some benefits in *P. euramericana* growth performance compared to the biochar that, both alone and in combination with the compost, seemed to have neutral or antagonistic effects on plant morphological and physiological traits, although with some differences during the diverse plant growth phases.

In detail, during the first months of plant growth (T0–T5), the biochar or compost addition alone had a negligible or negative effect on *P. euramericana* growth and development. Conversely, after the winter dormancy phase (T9–T12), the situation was quite different, and their addition led to an increase in plant height and in leaf number reaching the maximum in the compost alone amended substrate. The study by Jarvis et al. [77] also showed that the addition of compost soil conditioner significantly increased the height growth of alder, birch, and aspen species. Similar to the results of the present study, Heiskanen et al. [78] found that compost increased the growth of pine, willow, red clover, red fescue, and maiden pink seedlings. This positive effect was also confirmed by phenological analysis, which showed that the two amendments alone accelerated poplar growth and development compared to the control, leading to 60% more seedlings having buds with fully formed leaves after dormancy breaking. Nevertheless, the compost and biochar mixture induced slowing in bud development and overall plant growth.

The time effect could have been due to the fact that *P. euramericana* plants were potted as cuttings, and it was possible that nutrients in initial cuttings were translocated to offset nutrient demand. Successively, the ability of these two amendments to ameliorate substrate characteristics could have increased the available nutrient amounts for plants and, consequently, accelerated plant growth and development [77,78]. Indeed, despite the addition of biochar or compost alone having no effect on soil pH [20,79], they increased soil electrical conductivity and cation exchange capacity; moreover, both amendments ameliorated substrate chemical characteristics by increasing nutrient concentrations (TOC and Ntot, and Ptot and Pav), probably due to their content of organic matter and inorganic ions (e.g., N and P) [80]. Furthermore, the biochar amendment could have facilitated the biochemical cycling of nitrogen and phosphorus [27], and the compost could have improved soil aeration and water availability [78].

If growth responses to the biochar and compost alone addition were mainly due to nutrient provision, their combined use negatively affected the growth of all plant aboveground parts, despite enhanced soil nutrient availability. Seehausen et al. [81] also found that the combined addition of biochar and compost in nutrient-limited media had neutral or antagonistic interactive effects on several plant growth traits and physiological performance. More specifically, the authors found mostly neutral effects on *Salix purpurea* plants and antagonistic effects on maximum leaf area, aboveground, and belowground biomass, reproductive allocation, maximum plant height, chlorophyll fluorescence, and stomatal conductance of *Abutilon theophrasti* plants. This could be related to a saturation of plant nutrient demands resulting in a non-additive positive effect [82] and might have brought to a decline in poplar growth [83]. The biochar and compost combination might also have resulted in an oversupply of toxic elements (e.g., Al, As, Pb) and micronutrients (e.g., B, Mn) [81]. Alternatively, the poplar growth decline may also be due to the fact that biochar can immobilize/retain nutrients [23] thus reacting with them and acting as a competitor instead of providing nutrients to plants. For example, biochar could facilitate phosphate precipitation/absorption reactions and lead to reduced P and N availability to plants [84,85], as observed in the results reported above. Indeed, biochar has a high cation exchange capacity, which may significantly increase nutrient retention because of a high surface charge [78].

In the present study, the soil–sand mix used as the growing plant substrate was already moderately subalkaline, and an improvement in pH after biochar application was not expected, as also reported in other studies [86,87]. It is well documented that biochar application impacts soil pH, which, together with cation exchange capacity, influences nutrient interactions in soil. Given that, in the present study, biochar application did not affect substrate pH, immobilization, and retainment of nutrients were likely to be attributed to the increases in cation exchange capacity. Similarly, Liu et al. [88] documented that adsorption of NH_4^+-N may be due to the high cation exchange capacity of biochar, and adsorption of ammonium by biochar has also been well documented by Spokas et al. [89]. Nguyen et al. [90] demonstrated, in a meta-analysis, that biochar addition reduced soil inorganic nitrogen by about 11% (NH_4^+) and 10% (NO_3^-) per 56 works published between 2010 and 2015. The soil availability of other important micronutrients can be affected by biochar amendments. Sadowska et al. [91] reported that significantly more soluble Ca, K, and SO_4^{2-} were found in the soil amended with biochar (pH 7.03) as compared with the control. Similarly, Bista et al. [92] and Marks et al. [93] found that biochar increased soil concentrations of K^+ and SO_4^{2-}, which was attributed to a direct additive effect. Nevertheless, again, the availability of nutrients and micronutrients is influenced by biochar and soil properties.

As mentioned, many of the changes in nutrient cycling are related to specific biochar characteristics (e.g., feedstock, pyrolysis temperature), as well as to how it ages within soil [10]. For instance, biochar has an inhibitory effect on soil aging, and the intermittent addition of fresh biochar biomass may be necessary for optimal nutrient cycling in soil [23]. Furthermore, biochar produced from plant feedstocks generally decomposes faster than biochar produced from wood or grasses [94]. Therefore, the fact that our biochar was produced from orchid pruning biomass could explain its relatively fast decomposition rate, which may have led to the lack of positive and long-lasting effects on plant growth. Indeed, as reported in Trupiano et al. [53] about physicochemical characteristics of the soil amendments used in the present study, the C/N ratio for the biochar was high (125.5), and this may confirm a biochar fast decomposition rate. In addition, soil clay particles have been found to play an important role in biochar stabilization [95]. Thus, the relatively low clay content in the growing substrates may have provided low protection against biochar biophysical degradation.

Another result to point out was that the ability of the biochar and/or compost amendment to induce growth acceleration was not reflected in biomass accumulation. In detail, we found a reduction in the stem biomass of *P. euramericana* grown on all the amended substrates (B, Co, and BCo) and in the leaf and root biomass of plant grown on the biochar and compost mix. Moreover, we also found a higher S/R ratio in plants grown on the substrate amended with both biochar and compost. The mixture also negatively affected the very fine and fine root length, contrary to the compost addition alone, which showed a great positive impact on this parameter. The results were presumably due to the biochar, which remained ineffective in other experiments [96]. Mertens et al. [96] showed that root length density, fine root dry weight, shoot dry weight, and shoot-root ratio of three-year-old *Spondias tuberosa* seedlings grown on soil amended with biochar did not report significant differences compared to control soil.

Shoot/root biomass partitioning is an important mechanism by which plants cope with the limitations imposed by growth-constraining resources in the environment [97]. Song et al. [98] indicated that nutrient deficiency (nitrogen and potassium) promoted root growth and increased N and K allocation in storage organs, especially promoting the growth of fine roots. In addition, improving fertilizer under sufficient nutrient conditions did not promote nutrient accumulation in the storage organs, and most of the nutrients were lost with defoliation [98]. In our case, the high value of S/R ratio for poplar plants grown in the biochar-compost mixture confirmed the findings reported above. As previously explained, the combined use of the two amendments might have led to an accumulation of micronutrients, which, in accordance with the study of Song et al. [98], did not promote root growth that could be potentially related also to a high vulnerability to water deficiency [99].

Furthermore, taking into account the fact that plant investment in fine root production is related to efficiently spending resources on water and nutrient uptake [100], the raised concentrations of all nutrient contents in the biochar and compost mix (BCo) should be the cause of very fine and fine root length decreasing [101]. Fine roots are a highly dynamic part of tree biomass that not only have a large influence on forest water and nutrient cycles, but also represent a major source of soil organic carbon [21,64]. The morphological plasticity of roots has been reported in previous studies, especially for fine root fractions [102,103]. In light of this knowledge, we may assert that poplar plants, when growing in substrates amended by the biochar or compost alone, optimized the investment of carbon to dynamically and plastically change root morphology and ensured water and nutrient uptake, without affecting root biomass accumulation [104]. In particular, the compost addition alone enhanced the very fine and fine root length, presumably, for the low soil moisture content, high nutrient levels, and good soil aeration [77,79]; whereas, the compost and biochar combination seemed to confirm an antagonistic effect also on root system development, inducing a decrease in both root biomass and very fine and fine root length.

Plant growth and biomass accumulation are tightly coordinated with photosynthesis to meet the plant demand for the energy required during development [46,47]. Biomass change is the mass balance between production and loss. However, in this counting, it is important to take into account that biomass can be allocated to growth, defense, partitioning, or storage [105], which participate differently in *sensu stricto* or *sensu lato* plant growth. Indeed, plant growth in *sensu stricto* refers to the irreversible increase in total biomass stored as compounds that form the structure of plants (cellulose and lignin in the cell walls, lipids in the membranes, and proteins within the cell). Growth in *sensu lato* refers to irreversible increase in cell, organ, or plant volume, together with a reversible increase of storage proteins and lipids, secondary metabolites, or nonstructural carbohydrates (e.g., sugars, starch). Storage compounds can be reversible depleted (negative biomass change) for the production, accumulation, and release of secondary metabolites as a buffer against any disturbance-based fluctuation in carbohydrate assimilation, particularly important for plant defense. Thus, structural growth and storage are potentially competing for carbon resources to ensure the growth–defense tradeoff [106] and must therefore be highly regulated.

On the basis of these findings, the accelerated growth rate induced by the biochar could be related to changes in the priority of soluble carbohydrate distribution to growing organs [94] that should be guaranteed by active photosynthetic activity (high leaf chlorophyll content) in plants grown on biochar amended substrates (biochar alone and in combination with the compost).

Several studies showed that exogenous application of biochar increased the chlorophyll content due to the enhanced availability of nutrients and water [107,108] and, contrary to that revealed here, is generally strictly related to a higher plant biomass accumulation. Other studies demonstrated that the addition of biochar could misbalance the photosynthetic machinery and impair the mechanisms recognizing pathogen-derived molecules inducing plant defense machinery dysfunction [109,110]. Thus, we can hypothesize, in the case of poplar plants, a diversion of resources away from energy reserves and toward defense to optimize plant fitness [111].

In particular, considering that in poplar grown on the biochar amended substrates (biochar alone and in combination with compost), the root lignin content increased in accordance with the high leaf chlorophyll content, the carbon flux could be extended in root toward the phenylpropanoid pathway, and possibly includes the biosynthesis of lignin [112,113]. It has recently been reported that biochar induces the up-regulation of several enzymes involved in lignin synthesis that should be indispensable for plant growth-defense tradeoff, acting as an important physical barrier that protects against pathogen invasion and preventing toxic compounds ingress [114–116]. However, to date, crosstalk among lignin content and plant growth–defense tradeoff is complex and remains unpredictable due to the limited understanding of the underlying mechanisms [117].

In conclusion, our study showed that biochar and/or compost applications improved growing medium physicochemical characteristics by increasing electrical conductivity, cation exchange capacity, and nutrient concentrations. These ameliorations led to accelerated *P. euramericana* growth and development—as revealed by poplar phenology evaluation—when the biochar and compost were used alone, whereas the biochar-compost combination induced a slowing in plant bud development. However, the amendment additions to the growing substrate had negligible or negative effects on poplar morphological traits, except for the compost added alone, which increased plant height and very fine and fine root length. The biochar–compost combination was found to have negative effects on plant growth that could have been due to an accumulation of nutrients and/or micronutrients that, at high concentrations, could become toxic and reduce plant development.

Overall, our results showed that the compost addition alone promoted *P. euramericana* growth without affecting the structural features and, thus, plant biomass accumulation. However, the compost was able to enhance very fine and fine root lengths, guaranteeing water and nutrient uptake. Conversely, the biochar, both alone and in combination with the compost, produced a negative/summative effect that was reflected in a carbon metabolism shift (from primary to secondary) toward lignin biosynthesis for optimizing the growth-defense tradeoff.

Consequently, the compost amendment alone should be the best solution for the nursery-produced poplar seedlings, being able to improve both substrate properties and root system characteristics, which are key aspects in a forest restoration program. These characteristics may make seedlings better able to cope with the period after transplanting, overcoming periods of aridity exacerbated by current climate change, and enhancing plant stability in steep soils.

Future work is anticipated on investigations for the use of biochar and compost in the long term, focusing on all rhizosphere components and functional interaction. Given the importance of nutrient concentration and availability in soil for plant growth, it is critical that future studies provide more comprehensive details on the surface properties of biochar in the soil environment, comparing biochar-nutrient interaction against biochar-compost. Moreover, collectively, our findings suggest that the responses of soil and plants to biochar are strongly influenced by amendment physicochemical characteristics. Therefore, determining the practical effectiveness of biochar in the long term requires additional studies on the type and rate of biochar application, in addition to optimization of feedstock properties and pyrolysis conditions suitable for better biochar usage. Including short-term and long-term evaluation of biochar must complement each other to unravel the possible effect of age on biochar, clarifying the effects of aged versus fresh biochar. It may also be important to evaluate biochar and compost both developed from the same feedstock as a part of future line of research. Additionally, a better understanding of the factors determining very fine and fine root lifespan, turnover, and decomposition will be crucial for a mechanistic insight into tree responses to diverse amendments and changing environments and for the quantification of forest carbon turnover. This information will be useful to scientists/managers involved in smart selection and innovation in properly applying soil amendments in effective forest restoration strategies.

Supplementary Materials: The following supporting information can be downloaded at: https://www.mdpi.com/article/10.3390/f13040550/s1, Table S1: Seasonal changes during experimental trial.

Author Contributions: Conceptualization, D.C., G.S.S. and D.T.; methodology and formal analysis, E.D.Z., A.M., M.T. and D.T.; software, validation and data curation, M.S.; writing—original draft preparation, M.S. and D.T.; writing—review and editing, M.S., E.D.Z., A.M., M.T., D.C., G.S.S. and D.T.; supervision, D.C., G.S.S. and D.T. All authors have read and agreed to the published version of the manuscript.

Funding: This research received no external funding.

Data Availability Statement: Not applicable.

Conflicts of Interest: The authors declare no conflict of interest.

References

1. Li, Y.; Ye, Y.; Fang, X.; Liu, Y. Reconstruction of agriculture-driven deforestation in western Hunan province of China during the 18th century. *Land* **2022**, *11*, 181. [CrossRef]
2. Skole, D.L.; Mbow, C.; Mugabowindekwe, M.; Brandt, M.S.; Samek, J.H. Trees outside of forests as natural climate solutions. *Nat. Clim. Chang.* **2021**, *11*, 1013–1016. [CrossRef]
3. Hoekman, D.; Kooij, B.; Quiñones, M.; Vellekoop, S.; Carolita, I.; Budhiman, S.; Arief, R.; Roswintiarti, O. Wide-area near-real-time monitoring of tropical forest degradation and deforestation using Sentinel-1. *Remote Sens.* **2020**, *12*, 3263. [CrossRef]
4. Curtis, P.G.; Slay, C.M.; Harris, N.L.; Tyukavina, A.; Hansen, M.C. Classifying drivers of global forest loss. *Science* **2018**, *361*, 1108–1111. [CrossRef]
5. Ritchie, H.; Roser, M. Forests and Deforestation. 2021. Available online: https://ourworldindata.org/forests-and-deforestation (accessed on 20 January 2022).
6. Breshears, D.D.; Fontaine, J.B.; Ruthrof, K.X.; Field, J.P.; Feng, X.; Burger, J.R.; Law, D.J.; Kala, J.; Hardy, G.E.S.J. Underappreciated plant vulnerabilities to heat waves. *New Phytol.* **2021**, *231*, 32–39. [CrossRef]
7. Matusick, G.; Ruthrof, K.X.; Brouwers, N.C.; Dell, B.; Hardy, G.S. Sudden forest canopy collapse corresponding with extreme drought and heat in a Mediterranean-type eucalypt forest in south western Australia. *Eur. J. For. Res.* **2013**, *132*, 497–510. [CrossRef]
8. Payne, R.J.; Anderson, A.R.; Sloan, T.; Gilbert, P.; Newton, A.; Ratcliffe, J.; Mauquoy, D.; Jessop, W.; Andersen, R. The future of peatland forestry in Scotland: Balancing economics, carbon and biodiversity. *Scott. For.* **2018**, *100*, 34–40.
9. Brown, I. Challenges in delivering climate change policy through land use targets for afforestation and peatland restoration. *Environ. Sci. Policy* **2020**, *107*, 36–45. [CrossRef]
10. Dumroese, R.K.; Landis, T.D.; Pinto, J.R.; Haase, D.L.; Wilkinson, K.W.; Davis, A.S. Meeting forest restoration challenges: Using the target plant concept. *Reforesta* **2016**, *1*, 37–52. [CrossRef]
11. Mansourian, S. In the eye of the beholder: Reconciling interpretations of forest landscape restoration. *Land Degrad. Dev.* **2018**, *29*, 2888–2898. [CrossRef]
12. Le, H.D.; Smith, C.; Herbohn, J. Identifying interactions among reforestation success drivers: A case study from the Philippines. *Ecol. Modell.* **2015**, *316*, 62–77. [CrossRef]

13. Fargione, J.; Haase, D.L.; Burney, O.T.; Kildisheva, O.A.; Edge, G.; Cook-Patton, S.C.; Chapman, T.; Rempel, A.; Hurteau, M.D.; Davis, K.T.; et al. Challenges to the reforestation pipeline in the United States. *Front. For. Glob. Chang.* **2021**, *4*, 8. [CrossRef]
14. North, M.P.; Stevens, J.T.; Greene, D.F.; Coppoletta, M.; Knapp, E.E.; Latimer, A.M.; Restaino, C.M.; Tompkins, R.E.; Welch, K.R.; York, R.A.; et al. Tamm Review: Reforestation for resilience in dry western US forests. *For. Ecol. Manag.* **2019**, *432*, 209–224. [CrossRef]
15. Puettmann, M.; Sahoo, K.; Wilson, K.; Oneil, E. Life cycle assessment of biochar produced from forest residues using portable systems. *J. Clean. Prod.* **2020**, *250*, 119564. [CrossRef]
16. Davis, A.S.; Pinto, J.R. The scientific basis of the Target Plant Concept: An overview. *Forests* **2021**, *12*, 1293. [CrossRef]
17. Andivia, E.; Villar-Salvador, P.; Oliet, J.A.; Puértolas, J.; Dumroese, R.K. How can my research paper be useful for future meta-analyses on forest restoration plantations? *New For.* **2019**, *50*, 255–266. [CrossRef]
18. Meli, P.; Isernhagen, I.; Brancalion, P.H.; Isernhagen, E.C.; Behling, M.; Rodrigues, R.R. Optimizing seeding density of fast-growing native trees for restoring the Brazilian Atlantic Forest. *Restor. Ecol.* **2018**, *26*, 212–219. [CrossRef]
19. Brancalion, P.H.; Holl, K.D. Guidance for successful tree planting initiatives. *J. Appl. Ecol.* **2020**, *57*, 2349–2361. [CrossRef]
20. Drake, J.A.; Carrucan, A.; Jackson, W.R.; Cavagnaro, T.R.; Patti, A.F. Biochar application during reforestation alters species present and soil chemistry. *Sci. Total Environ.* **2015**, *514*, 359–365. [CrossRef]
21. Montagnoli, A.; Dumroese, R.K.; Terzaghi, M.; Pinto, J.R.; Fulgaro, N.; Scippa, G.S.; Chiatante, D. Tree seedling response to LED spectra: Implications for forest restoration. *Plant Biosyst.-Int. J. Deal. All Asp. Plant Biol.* **2018**, *152*, 515–523. [CrossRef]
22. Barrow, C. Biochar: Potential for countering land degradation and for improving agriculture. *Appl. Geogr.* **2012**, *34*, 21–28. [CrossRef]
23. Lefebvre, D.; Román-Dañobeytia, F.; Soete, J.; Cabanillas, F.; Corvera, R.; Ascorra, C.; Fernandez, L.E.; Silman, M. Biochar effects on two tropical tree species and its potential as a tool for reforestation. *Forests* **2019**, *10*, 678. [CrossRef]
24. Lehmann, J.; Rillig, M.; Thies, J.; Masiello, C.; Hockaday, W.; Crowley, D. Biochar effects on soil biota—A review. *Soil Biol. Biochem.* **2011**, *43*, 1812–1836. [CrossRef]
25. Woolf, D.; Amonette, J.E.; Street-Perrott, F.A.; Lehmann, J.; Joseph, S. Sustainable biochar to mitigate global climate change. *Nat. Commun.* **2010**, *1*, 56. [CrossRef] [PubMed]
26. Kameyama, K.; Miyamoto, T.; Iwata, Y.; Shiono, T. Influences of feedstock and pyrolysis temperature on the nitrate adsorption of biochar. *Soil Sci. Plant Nutr.* **2016**, *62*, 180–184. [CrossRef]
27. Kavitha, B.; Reddy, P.V.L.; Kim, B.; Lee, S.S.; Pandey, S.K.; Kim, K.H. Benefits and limitations of biochar amendment in agricultural soils: A review. *J. Environ. Manag.* **2018**, *227*, 146–154. [CrossRef]
28. Thomas, S.C.; Gale, N. Biochar and forest restoration: A review and meta-analysis of tree growth responses. *New For.* **2015**, *46*, 931–946. [CrossRef]
29. Karim, M.; Halim, M.A.; Gale, N.V.; Thomas, S.C. Biochar effects on soil physiochemical properties in degraded managed ecosystems in northeastern Bangladesh. *Soil Syst.* **2020**, *4*, 69. [CrossRef]
30. Wu, S.; He, H.; Inthapanya, X.; Yang, C.; Lu, L.; Zeng, G.; Han, Z. Role of biochar on composting of organic wastes and remediation of contaminated soils—A review. *Environ. Sci. Pollut. Res.* **2017**, *24*, 16560–16577. [CrossRef]
31. Lal, R. Restoring soil quality to mitigate soil degradation. *Sustainability* **2015**, *7*, 5875–5895. [CrossRef]
32. Saer, A.; Lansing, S.; Davitt, N.H.; Graves, R.E. Life cycle assessment of a food waste composting system: Environmental impact hotspots. *J. Clean. Prod.* **2013**, *52*, 234–244. [CrossRef]
33. Post, W.M.; Kwon, K.C. Soil carbon sequestration and land-use change: Processes and potential. *Glob. Chang. Biol.* **2000**, *6*, 317–327. [CrossRef]
34. Diacono, M.; Montemurro, F. Long-term effects of organic amendments on soil fertility: A review. *Agron. Sustain. Dev.* **2010**, *30*, 401–422. [CrossRef]
35. Persiani, A.; Diacono, M.; Monteforte, A.; Montemurro, F. Agronomic performance, energy analysis and carbon balance comparing different fertilization strategies in horticulture under Mediterranean conditions. *Environ. Sci. Pollut. Res.* **2019**, *26*, 19250–19260. [CrossRef] [PubMed]
36. Pane, C.; Palese, A.M.; Spaccini, R.; Piccolo, A.; Celano, G.; Zaccardelli, M. Enhancing sustainability of a processing tomato cultivation system by using bioactive compost teas. *Sci. Hort.* **2016**, *202*, 117–124. [CrossRef]
37. Martínez-Blanco, J.; Muñoz, P.; Anton, A.; Rieradevall, J. Life cycle assessment of the use of compost from municipal organic waste for fertilization of tomato crops. *Resour. Conserv. Recycl.* **2009**, *53*, 340–351. [CrossRef]
38. Steiner, C.; Das, K.C.; Melear, N.; Lakly, D. Reducing nitrogen loss during poultry litter composting using biochar. *J. Environ. Qual.* **2010**, *39*, 1236–1242. [CrossRef]
39. Fischer, D.; Glaser, B. Synergisms between compost and biochar for sustainable soil amelioration. In *Management of Organic Waste*; Kumar, S., Ed.; InTech: Rijeka, Croatia, 2012; pp. 167–198.
40. Liu, J.; Schulz, H.; Brandl, S.; Miehtke, H.; Huwe, B.; Glaser, B. Short-Term effect of biochar and compost on soil fertility and water status of a Dystric Cambisol in NE Germany under field conditions. *J. Plant Nutr. Soil Sci.* **2012**, *175*, 698–707. [CrossRef]
41. Grossnickle, S.C.; MacDonald, J.E. Seedling quality: History, application, and plant attributes. *Forests* **2018**, *9*, 283. [CrossRef]
42. Grossnickle, S.C. Why seedlings survive: Importance of plant attributes. *New For.* **2012**, *43*, 711–738. [CrossRef]
43. Grossnickle, S.C.; South, D.B. Seedling quality of southern pines: Influence of plant attributes. *Tree Plant. Notes* **2017**, *60*, 29–40.

44. Pinto, J.R. Morphology targets: What do seedling morphological attributes tell us. In *National Proceedings: Forest and Conservation Nursery Associations-2010*; RMRS-P-65; U.S. Department of Agriculture, Forest Service: Fort Collins, CO, USA, 2011; pp. 74–79.
45. Jha, K.K. Root carbon sequestration and its efficacy in forestry and agroforestry systems: A case of *Populus euramericana* I-214 cultivated in Mediterranean condition. *Not. Sci. Biol.* **2018**, *10*, 68–78. [CrossRef]
46. Fuertes, A.; Oliveira, N.; Cañellas, I.; Sixto, H.; Rodríguez-Soalleiro, R. An economic overview of *Populus* spp. in Short Rotation Coppice systems under Mediterranean conditions: An assessment tool for decision-making. *Renew. Sustain. Energy Rev.* **2021**, *151*, 111577. [CrossRef]
47. Pallardy, S.G.; Gibbins, D.E.; Rhoads, J.L. Biomass production by two-year-old poplar clones on floodplain sites in the Lower Midwest, USA. *Agrofor. Syst.* **2003**, *59*, 21–26. [CrossRef]
48. Nielsen, U.B.; Madsen, P.; Hansen, J.K.; Nord-Larsen, T.; Nielsen, A.T. Production potential of 36 poplar clones grown at medium length rotation in Denmark. *Biomass Bioenergy* **2014**, *64*, 99–109. [CrossRef]
49. Giri, A.; Kumar, G.; Arya, R.; Mishra, S.; Mishra, A.K. Carbon sequestration in *Populus deltoides* based agroforestry system in northern India. *Int. J. Chem. Stud.* **2019**, *7*, 2184–2188.
50. VanWallendael, A.; Lowry, D.B.; Hamilton, J.A. One hundred years into the study of ecotypes, new advances are being made through large-scale field experiments in perennial plant systems. *Curr. Plant Biol.* **2022**, *66*, 102152. [CrossRef] [PubMed]
51. U.S Department of Agriculture, Natural Resources Conservation Service. National Soil Survey Handbook. Available online: http://www.nrcs.usda.gov/wps/portal/nrcs/detail/soils/ref/?cid=nrcs142p2_054242 (accessed on 27 January 2022).
52. Alfano, G.; Lustrato, G.; Lima, G.; Vitullo, D.; Ranalli, G. Characterization of composted olive mill wastes to predict potential plant disease suppressiveness. *Biol. Control* **2011**, *58*, 199–207. [CrossRef]
53. Trupiano, D.; Cocozza, C.; Baronti, S.; Amendola, C.; Vaccari, F.P.; Lustrato, G.; Di Lonardo, S.; Fantasma, F.; Tognetti, R.; Scippa, G.S. The effects of biochar and its combination with compost on lettuce (*Lactuca sativa* L.) growth, soil properties, and soil microbial activity and abundance. *Int. J. Agron.* **2017**, *12*, 3158207. [CrossRef]
54. Black, C.A. *Methods of Soil Analysis: Part I Physical and Mineralogical Properties*; American Society of Agronomy: Madison, WI, USA, 1965.
55. Rayment, G.E.; Higginson, F.R. *Australian Laboratory Handbook of Soil and Water Chemical Methods*; Inkata Press Pty Ltd.: Melbourne, Australia, 1992.
56. Jones, J.B., Jr.; Kalra, Y.P. Soil testing and plant analysis activities-The United States and Canada. *Commun. Soil Sci. Plant Anal.* **1992**, *23*, 2015–2027. [CrossRef]
57. Mehlich, A. Use of triethanolamine acetate-barium hydroxide buffer for the determination of some base exchange properties and lime requirement of soil. *Soil Sci. Soc. Am. Proc.* **1938**, *29*, 374–378. [CrossRef]
58. Dumas, J.B.A. Procédés de l'analyse organique. *Ann. Chim. Phys.* **1831**, *247*, 198–213.
59. Bowman, R.A. A rapid method to determine total phosphorus in soils. *Soil Sci. Soc. Am. J.* **1988**, *52*, 1301–1304. [CrossRef]
60. Olsen, S.R.; Cole, C.V.; Watanabe, F.S.; Dean, L.A. *Estimation of Available Phosphorus in Soils by Extraction with Sodium Bicarbonate. USDA Circular 939*; U.S. Government Publishing Office: Washington, DC, USA, 1954.
61. Bouyoucos, G.J. Hydrometer method improved for making particle size analysis of soils. *Agron. J.* **1962**, *54*, 464–465. [CrossRef]
62. Beretta, N.; Silbermann, A.V.; Paladino, L.; Torres, D.; Bassahun, D.; Musselli, R.; Lamohte, A.G. Soil texture analyses using a hydrometer modification of the Bouyoucos method. *Cienc. Investig. Agrar.* **2014**, *41*, 263–271. [CrossRef]
63. Kemper, W.D.; Koch, E.J. *Aggregate Stability of Soils from Western USA and Canada USDA Technical Bulletin No.1355*; US Government Printing Office: Washington, DC, USA, 1966.
64. Montagnoli, A.; Di Iorio, A.; Terzaghi, M.; Trupiano, D.; Scippa, G.S.; Chiatante, D. Influence of soil temperature and water content on fine-root seasonal growth of European beech natural forest in Southern Alps, Italy. *Eur. J. For. Res.* **2014**, *133*, 957–968. [CrossRef]
65. Trupiano, D.; Rocco, M.; Renzone, G.; Scaloni, A.; Montagnoli, A.; Terzaghi, M.; Di Iorio, A.; Chiatante, D.; Scippa, G.S. Poplar woody root proteome during the transition dormancy-active growth. *Plant Biosyst.-Int. J. Deal. All Asp. Plant Biol.* **2013**, *147*, 1095–1100. [CrossRef]
66. Doster, M.A.; Bostock, R.M. Quantification of lignin formation in almond bark in response to wounding and infection by *Phytophthora* species. *Phytopathology* **1988**, *784*, 73–477.
67. Trupiano, D.; Di Iorio, A.; Montagnoli, A.; Lasserre, B.; Rocco, M.; Grosso, A.; Scaloni, A.; Marra, M.; Chiatante, D.; Scippa, G.S. Involvement of lignin and hormones in the response of woody poplar taproots to mechanical stress. *Physiol. Plant.* **2012**, *146*, 39–52. [CrossRef]
68. Inskeep, W.P.; Bloom, P.R. Extinction coefficients of chlorophyll a and b in *N,N* Dimethylformamide and 80% acetone. *Plant Physiol.* **1985**, *77*, 483–485. [CrossRef]
69. Kroetsch, D.; Wang, C. Particle size distribution. *Soil Sampl. Methods Anal.* **2008**, *2*, 713–725.
70. Wolf, M.; Lehndorff, E.; Wiesenberg, G.L.B.; Stockhausen, M.; Schward, L.; Amelung, W. Towards reconstruction of past fire regimes from geochemical analysis of charcoal. *Org. Geochem.* **2013**, *55*, 11–21. [CrossRef]
71. Ippolito, J.A.; Spokas, K.A.; Novak, J.M.; Lentz, R.D.; Cantrell, K.B. Biochar elemental composition and factors influencing nutrient retention. In *Biochar for Environmental Management: Science, Technology and Implementation*, 2nd ed.; Lehmann, J., Joseph, S., Eds.; Routledge: New York, NY, USA, 2015; pp. 137–161.

72. Ippolito, J.A.; Cui, L.; Kammann, C.; Wrage-Mönnig, N.; Estavillo, J.M.; Fuertes-Mendizabal, T.; Cayuela, M.L.; Sigua, G.; Novak, J.; Spokas, K.; et al. Feedstock choice, pyrolysis temperature and type influence biochar characteristics: A comprehensive meta-data analysis review. *Biochar* **2020**, *2*, 421–438. [CrossRef]
73. Ippolito, J.A.; Berry, C.M.; Strawn, D.G.; Novak, J.M.; Levine, J.; Harley, A. Biochars reduce mine land soil bioavailable metals. *J. Environ. Qual.* **2017**, *46*, 411–419. [CrossRef]
74. Khan, N.; Clark, I.; Sánchez-Monedero, M.A.; Shea, S.; Meier, S.; Bolan, N. Maturity indices in co-composting of chicken manure and sawdust with biochar. *Bioresour. Technol.* **2014**, *168*, 245–251. [CrossRef]
75. Zhang, J.; Zeng, G.; Chen, Y.; Yu, M.; Yu, Z.; Li, H.; Yu, Y.; Huang, H. Effects of physico-chemical parameters on the bacterial and fungal communities during agricultural waste composting. *Bioresour. Technol.* **2011**, *102*, 2950–2956. [CrossRef]
76. Reyes-Torres, M.; Oviedo-Ocaña, E.R.; Dominguez, I.; Komilis, D.; Sánchez, A. A systematic review on the composting of green waste: Feedstock quality and optimization strategies. *Waste Manag.* **2018**, *77*, 486–499. [CrossRef]
77. Järvis, J.; Ivask, M.; Nei, L.; Kuu, A.; Luud, A. Effect of green waste compost application on afforestation success. *Balt. For.* **2016**, *22*, 90–97.
78. Heiskanen, J.; Hagner, M.; Ruhanen, H.; Mäkitalo, K. Addition of recyclable biochar, compost and fibre clay to the growth medium layer for the cover system of mine tailings: A bioassay in a greenhouse. *Environ. Earth Sci.* **2020**, *79*, 422. [CrossRef]
79. Lazdina, D.; Bardule, A.; Lazdins, A.; Stola, J. Use of waste water sludge and wood ash as fertiliser for *Salix* cultivation in acid peat soils. *Agron. Res.* **2011**, *9*, 305–314.
80. Gul, S.; Whalen, J.K. Biochemical cycling of nitrogen and phosphorus in biochar amended soils. *Soil Biol. Biochem.* **2016**, *103*, 1–15. [CrossRef]
81. Seehausen, M.L.; Gale, N.V.; Dranga, S.; Hudson, V.; Liu, N.; Michener, J.; Thurston, E.; Williams, C.; Smith, S.M.; Thomas, S.C. Is there a positive synergistic effect of biochar and compost soil amendments on plant growth and physiological performance? *Agronomy* **2017**, *7*, 13. [CrossRef]
82. Tessier, J.T.; Raynal, D.J. Use of nitrogen to phosphorus ratios in plant tissue as an indicator of nutrient limitation and nitrogen saturation. *J. Appl. Ecol.* **2003**, *40*, 523–534. [CrossRef]
83. Tisdale, S.L.; Nelson, W.L.; Beaton, J.D. *Soil Fertility and Fertilizers*; Collier Macmillan Publishers: London, UK, 1985.
84. Xu, G.; Zhang, Y.; Sun, J.; Shao, H. Negative interactive effects between biochar and phosphorus fertilization on phosphorus availability and plant yield in saline sodic soil. *Sci. Total Environ.* **2016**, *568*, 910–915. [CrossRef] [PubMed]
85. Joseph, S.; Kammann, C.I.; Shepherd, J.G.; Conte, P.; Schmidt, H.P.; Hagemann, N.; Rich, A.M.; Marjo, C.E.; Allen, J.; Munroe, P.; et al. Microstructural and associated chemical changes during the composting of a high temperature biochar: Mechanisms for nitrate, phosphate and other nutrient retention and release. *Sci. Total Environ.* **2018**, *618*, 1210–1223. [CrossRef]
86. Hossain, M.Z.; Bahar, M.M.; Sarkar, B.; Donne, S.W.; Ok, Y.S.; Palansooriya, K.N.; Kirkham, M.B.; Chowdhury, S.; Bolan, N. Biochar and its importance on nutrient dynamics in soil and plant. *Biochar* **2020**, *2*, 379–420. [CrossRef]
87. Biederman, L.A.; Harpole, W.S. Biochar and its effects on plant productivity and nutrient cycling: A meta-analysis. *GCB Bioenergy* **2013**, *5*, 202–214. [CrossRef]
88. Liu, Z.; He, T.; Cao, T.; Yang, T.; Meng, J.; Chen, W. Effects of biochar application on nitrogen leaching, ammonia volatilization and nitrogen use efficiency in two distinct soils. *J. Soil Sci. Plant Nutr.* **2017**, *17*, 515–528. [CrossRef]
89. Spokas, K.A.; Novak, J.M.; Venterea, R.T. Biochar's role as an alternative N-fertilizer: Ammonia capture. *Plant Soil* **2012**, *350*, 35–42. [CrossRef]
90. Nguyen, T.T.N.; Xu, C.Y.; Tahmasbian, I.; Che, R.; Xu, Z.; Zhou, X.; Wallace, H.M.; Bai, S.H. Effects of biochar on soil available inorganic nitrogen: A review and metaanalysis. *Geoderma* **2017**, *288*, 79–96. [CrossRef]
91. Sadowska, U.; Domagała-Świątkiewicz, I.; Żabiński, A. Biochar and its effects on plant–soil macronutrient cycling during a three-year field trial on sandy soil with peppermint (*Mentha piperita* L.). Part I: Yield and macro element content in soil and plant biomass. *Agronomy* **2020**, *10*, 1950. [CrossRef]
92. Bista, P.; Ghimire, R.; Machado, S.; Pritchett, L. Biochar effects on soil properties and wheat biomass vary with fertility management. *Agronomy* **2019**, *9*, 623. [CrossRef]
93. Marks, A.N.; Mattan, S.; Alcañiz, J.M.; Pérez-Herrero, E.; Domene, X. Gasifier biochar effects on nutrient availability, organic matter mineralization, and soil fauna activity in a multi-year Mediterranean trial. *Agric. Ecosyst. Environ.* **2016**, *215*, 30–39. [CrossRef]
94. Wang, J.; Xiong, Z.; Kuzyakov, Y. Biochar stability in soil: Metaanalysis of decomposition and priming effects. *Glob. Chang. Biol. Bioenergy* **2016**, *8*, 512–523. [CrossRef]
95. Ventura, M.; Alberti, G.; Panzacchi, P.; Vedove, G.D.; Miglietta, F.; Tonon, G. Biochar mineralization and priming effect in a poplar short rotation coppice from a 3-year field experiment. *Biol. Fertil. Soils* **2019**, *55*, 67–78. [CrossRef]
96. Mertens, J.; Germer, S.; Germer, J.; Sauerborn, J. Comparison of soil amendments for reforestation with a native multipurpose tree under semiarid climate: Root and root tuber response of *Spondias tuberosa*. *For. Ecol. Manag.* **2017**, *396*, 1–10. [CrossRef]
97. Mašková, T.; Herben, T. Root: Shoot ratio in developing seedlings: How seedlings change their allocation in response to seed mass and ambient nutrient supply. *Ecol. Evol.* **2018**, *8*, 7143–7150. [CrossRef]
98. Song, X.; Wan, F.; Chang, X.; Zhang, J.; Sun, M.; Liu, Y. Effects of nutrient deficiency on root morphology and nutrient allocation in *Pistacia chinensis* Bunge seedlings. *Forests* **2019**, *10*, 1035. [CrossRef]

99. Jordan-Meille, L.; Martineau, E.; Bornot, Y.; Lavres, J.; Abreu-Junior, C.H.; Domec, J.C. How does water-stressed corn respond to potassium nutrition? A shoot-root scale approach study under controlled conditions. *Agriculture* **2018**, *8*, 180. [CrossRef]
100. West, J.B.; Espeleta, J.F.; Donovan, L.A. Fine root production and turnover across a complex edaphic gradient of a *Pinus palustris*–Aristida stricta savanna ecosystem. *For. Ecol. Manag.* **2004**, *189*, 397–406. [CrossRef]
101. Cheng, Y.; Han, Y.; Wang, Q.; Wang, Z. Seasonal dynamics of fine root biomass, root length density, specific root length, and soil resource availability in a *Larix gmelinii* plantation. *Front. Biol. China* **2006**, *1*, 310–317. [CrossRef]
102. Bjork, R.G.; Majdi, M.; Klemedtsson, L.; Jonsson, L.L.; Molau, U. Long-term warming effects on root morphology, root mass distribution, and microbial activity in two dry tundra plant communities in northern Sweden. *New Phytol.* **2007**, *176*, 862–873. [CrossRef] [PubMed]
103. Makita, N.; Hirano, Y.; Mizoguchi, T.; Kominami, Y.; Dannoura, M.; Ishii, H.; Finer, L.; Kanazawa, Y. Very fine roots respond to soil depth: Biomass allocation, morphology, and physiology in a broad-leaved temperate forest. *Ecol. Res.* **2011**, *26*, 95–104. [CrossRef]
104. Amendola, C.; Montagnoli, A.; Terzaghi, M.; Trupiano, D.; Oliva, F.; Baronti, S.; Miglietta, F.; Chiatante, D.; Scippa, G.S. Short-term effects of biochar on grapevine fine root dynamics and arbuscular mycorrhizae production. *Agric. Ecosyst. Environ.* **2017**, *239*, 236–245. [CrossRef]
105. Hilty, J.; Muller, B.; Pantin, F.; Leuzinger, S. Tansley review: Plant growth: The what, the how, and the why. *New Phytol.* **2021**, *232*, 25–41. [CrossRef]
106. Lind, E.M.; Borer, E.; Seabloom, E.; Adler, P.; Bakker, J.D.; Blumenthal, D.M.; Crawley, M.; Davies, K.; Firn, J.; Gruner, D.S.; et al. Life-history constraints in grassland plant species: A growth-defence trade-off is the norm. *Ecol. Lett.* **2013**, *16*, 513–521. [CrossRef]
107. Agegnehu, G.; Bass, A.M.; Nelson, P.N.; Bird, M.I. Benefits of biochar, compost and biochar–compost for soil quality, maize yield and greenhouse gas emissions in a tropical agricultural soil. *Sci. Total Environ.* **2016**, *543*, 295–306. [CrossRef]
108. Zhang, D.; Pan, G.; Wu, G.; Kibue, G.W.; Li, L.; Zhang, X.; Zheng, J.; Zheng, J.; Cheng, K.; Joseph, S. Biochar helps enhance maize productivity and reduce greenhouse gas emissions under balanced fertilization in a rainfed low fertility inceptisol. *Chemosphere* **2016**, *142*, 106–113. [CrossRef]
109. Viger, M.; Hancock, R.D.; Miglietta, F.; Taylor, G. More plant growth but less plant defence? First global gene expression data for plants grown in soil amended with biochar. *Glob. Chang. Biol. Bioenergy* **2015**, *7*, 658–672. [CrossRef]
110. Polzella, A.; De Zio, E.; Arena, S.; Scippa, G.S.; Scaloni, A.; Montagnoli, A.; Chiatante, D.; Trupiano, D. Toward an understanding of mechanisms regulating plant response to biochar application. *Plant Biosyst.* **2019**, *153*, 163–172. [CrossRef]
111. Simiele, M.; Sferra, G.; Lebrun, M.; Renzone, G.; Bourgerie, S.; Scippa, G.S.; Morabito, D.; Scaloni, A.; Trupiano, D. In-depth study to decipher mechanisms underlying *Arabidopsis thaliana* tolerance to metal (loid) soil contamination in association with biochar and/or bacteria. *Environ. Exp. Bot.* **2021**, *182*, 104335. [CrossRef]
112. Amthor, J.S. Efficiency of lignin biosynthesis: A quantitative analysis. *Ann. Bot.* **2003**, *91*, 673–695. [CrossRef] [PubMed]
113. Rogers, L.A.; Campbell, M.M. The genetic control of lignin deposition during plant growth and development. *New Phytol.* **2004**, *164*, 17–30. [CrossRef] [PubMed]
114. Ithal, N.; Recknor, J.; Nettleton, D.; Maier, T.; Baum, T.J.; Mitchum, M.G. Developmental transcript profiling of cyst nematode feeding cells in soybean roots. *Mol. Plant Microbe Interact.* **2007**, *20*, 510–525. [CrossRef] [PubMed]
115. Vanholme, R.; Storme, V.; Vanholme, B.; Sundin, L.; Christensen, J.H.; Goeminne, G.; Halpin, C.; Rohde, A.; Morreel, K.; Boerjan, W. A systems biology view of responses to lignin biosynthesis perturbations in *Arabidopsis*. *Plant Cell* **2012**, *24*, 3506–3529. [CrossRef]
116. Jaiswal, A.K.; Alkan, N.; Elad, Y.; Sela, N.; Philosoph, A.M.; Graber, E.R.; Frenkel, O. Molecular insights into biochar-mediated plant growth promotion and systemic resistance in tomato against *Fusarium* crown and root rot disease. *Sci. Rep.* **2020**, *10*, 13934. [CrossRef]
117. Liu, Q.; Luo, L.; Zheng, L. Lignins: Biosynthesis and biological functions in plants. *Int. J. Mol. Sci.* **2018**, *19*, 335. [CrossRef]

Article

Reconstruction of Conifer Root Systems Mapped with Point Cloud Data Obtained by 3D Laser Scanning Compared with Manual Measurement

Chikage Todo [1,2,*], Hidetoshi Ikeno [3], Keitaro Yamase [1], Toko Tanikawa [4], Mizue Ohashi [5], Masako Dannoura [6], Toshifumi Kimura [5] and Yasuhiro Hirano [2]

1 Hyogo Prefectural Technology Center for Agriculture, Forestry and Fisheries, Shiso 671-2515, Japan; kei-yamase@nike.eonet.ne.jp
2 Graduate School of Environmental Studies, Nagoya University, Nagoya 464-8601, Japan; yhirano@nagoya-u.jp
3 Faculty of Informatics, The University of Fukuchiyama, Fukuchiyama 620-0886, Japan; ikeno-hidetoshi@fukuchiyama.ac.jp
4 Graduate School of Bioagricultural Sciences, Nagoya University, Nagoya 464-8601, Japan; toko105@agr.nagoya-u.ac.jp
5 School of Human Science and Environment, University of Hyogo, Himeji 670-0092, Japan; ohashi@shse.u-hyogo.ac.jp (M.O.); kimura@shse.u-hyogo.ac.jp (T.K.)
6 Graduate School of Agriculture, Kyoto University, Kyoto 606-8502, Japan; dannoura.masako.4w@kyoto-u.ac.jp
* Correspondence: chikage_toudou@pref.hyogo.lg.jp

Abstract: Three-dimensional (3D) root system architecture (RSA) is a predominant factor in anchorage failure in trees. Only a few studies have used 3D laser scanners to evaluate RSA, but they do not check the accuracy of measurements. 3D laser scanners can quickly obtain RSA data, but the data are collected as a point cloud with a large number of points representing surfaces. The point cloud data must be converted into a set of interconnected axes and segments to compute the root system traits. The purposes of this study were: (i) to propose a new method for easily obtaining root point data as 3D coordinates and root diameters from point cloud data acquired by 3D laser scanner measurement; and (ii) to compare the accuracy of the data from main roots with intensive manual measurement. We scanned the excavated root systems of two *Pinus thunbergii* Parl. trees using a 3D laser scanner and neuTube software, which was developed for reconstructing the neuronal structure, to convert the point cloud data into root point data for reconstructing RSA. The reconstruction and traits of the RSA calculated from point cloud data were similar in accuracy to intensive manual measurements. Roots larger than 7 mm in diameter were accurately measured by the 3D laser scanner measurement. In the proposed method, the root point data were connected as a frustum of cones, so the reconstructed RSAs were simpler than the 3D root surfaces. However, the frustum of cones still showed the main coarse root segments correctly. We concluded that the proposed method could be applied to reconstruct the RSA and calculate traits using point cloud data of the root system, on the condition that it was possible to model both the stump and ovality of root sections.

Keywords: anchorage; coarse root; measurement method; *Pinus thunbergii*; root cross-sectional area; root system architecture

Citation: Todo, C.; Ikeno, H.; Yamase, K.; Tanikawa, T.; Ohashi, M.; Dannoura, M.; Kimura, T.; Hirano, Y. Reconstruction of Conifer Root Systems Mapped with Point Cloud Data Obtained by 3D Laser Scanning Compared with Manual Measurement. *Forests* **2021**, *12*, 1117. http://doi.org/10.3390/f12081117

Academic Editor: Antonio Montagnoli

Received: 14 July 2021
Accepted: 18 August 2021
Published: 21 August 2021

1. Introduction

Root systems anchor trees, capture and store resources, and sense the environment [1]. They contribute to the resistance to uprooting and prevent soil erosion, thus supporting slope stability [2,3].

Three-dimensional (3D) root system architecture (RSA) is a predominant factor in anchorage failure [4–8]. Root system traits that can be obtained from RSA include: the total root cross-sectional area (CSA) in a cross-section parallel to the slope [9], the pull-out resistance of roots as calculated from root diameter [10], and root volume in the root compartments such as tap root or horizontal roots [11]. Danjon et al. [7] reviewed the relationships between the RSA and anchorage and indicated the importance of compartment classification of RSA with different root system types by architectural analysis. Root systems must be excavated for such investigations, and given their size; it is necessary to obtain data efficiently.

The RSA of forest trees [12] is often calculated from 3D point data with root connections determined from *xyz* coordinates (ground surface position and depth) and diameter. Danjon and Reubens [12] reviewed and categorized methods for acquiring 3D root point data into manual, semi-automatic, and automatic methods. The manual measurement was the simplest but was labor and time-intensive [12,13]. Henderson et al. [14] determined the x and y coordinates by moving a T-square on an aluminum frame and the z coordinate by lowering a plumb bob and obtained data on the RSA of *Picea sitchensis* (Bong.) Carr. with a stem diameter at a breast height (DBH) of 9.0–11.5 cm. Mulatya et al. [15] obtained data on RSA by creating a level grid around *Melia volkensii* Gürke to measure coordinates where the branching and angles of the root system changed. Semi-automatic methods included the measurement of 3D root point data using a digital compass and an inclinometer [16] or a contact digitizer, and AMAPmod software with MTG coding [17–19]. Saint Cast et al. [20] used semi-automatic methods to obtain RSA data of *Pinus pinaster* Ait. from the seed to the mature stage and modeled the 3D structure of the entire root system. The method of using a 3D digitizer and yielding MTG format files was used extensively because it provided a precise 3D database of root architecture as a set of axes and segments [12], allowing the computation of many root traits. In addition, classifying segments in several root compartments, such as the zone of rapid taper (ZRT), has become the mainstream of RSA data analysis in recent years [12].

On the other hand, automatic methods that can obtain data on RSA rapidly collect data as a point cloud that consists of a very large number of points representing the root surfaces. Lontoc-Roy et al. [21] used computed tomography to obtain point cloud data of the root systems of tree seedlings, but this method cannot be used on large trees in the field. Additionally, the method did not convert point cloud data to root axes and segments. An automatic method that can be applied to large trees uses three-dimensional (3D) laser scanners. Although 3D laser scanners are expensive, they can collect point cloud data from 3D surfaces with high accuracy without contact, and thus, are used for measuring the aboveground parts of trees [22–26]. Lau et al. [27] reported the accuracy of the traits of branches > 10 cm in diameter reconstructed from the 3D quantitative structure models (QSM) from point cloud data using a 3D laser scanner in 4, up to 30 m high, tropical tree species with the leaves, scanned from the ground. However, only large branches (>40 cm in diameter) could be accurately reconstructed compared to manual measurements. Gärtner et al. [28] used 3D laser scanning to acquire point cloud data of the entire root system of 80-year-old *Picea abies* (L.) Karst. trees. Wagner et al. [29,30] created annual growth models at the root segments from 3D laser scans of the root system and 2D annual tree-ring data of 12-year-old *Pinus sylvestris* L. The mean absolute percentage error per volume of two root segments with diameters of 1.6 cm and 1.8 cm was reported to be 6.0% on average. Smith et al. [31] measured the volumes of 13 root systems of *P. abies* trees with a 19–47 cm stump diameter by converting point cloud data into polygon data using computer-aided design software. Only a few studies have used 3D laser scanning to evaluate root systems [12]. The 3D QSM reconstruction method has been used for both above- and below-ground of trees, but it requires specific parameters to fit the model, which could vary the accuracy of reconstruction [27,31]. Although only a theoretical comparison between semi-automatic methods and 3D laser scanning was reported in Danjon and Reubens [12], the accuracy of root point data reconstructed from 3D laser scan measurement has never been checked [31]. Moreover, 3D laser scanning has yet to be fully used for measuring the root system traits such as diameter and CSA from point cloud data.

Although it is the best available technique to describe the shapes of root surfaces, intensive intervention is still required for calculating root system traits [12]. Therefore, an automated step is needed to convert the point cloud data of the root surface measured by a 3D laser scanner to the root point data.

The purposes of this study were: (i) to propose a new method for easily obtaining root point data as 3D coordinates and root diameter from point cloud data acquired by a 3D laser scanner; and (ii) to compare the accuracy of the root point data with that collected by intensive manual measurement.

Interrelated root segments with just a base and an end diameter will be hereafter referred to as 'segments'. Interrelated root points, each with a diameter, forming axes, and segments, will be referred to as 'root point data'.

This study focused on *Pinus thunbergii* Parl., which is the main plantation in Japanese coastal forests. The excavated root systems of *P. thunbergii* were scanned using a 3D laser scanner, and the point cloud data were converted to root point data, from which we reconstructed the RSA data. Root point data were also acquired manually. We compared the RSA data obtained by the two methods and examined the reproducibility of both datasets from the root CSA.

2. Material and Methods

2.1. Test Trees

Root systems of *Pinus thunbergii* Parl. trees were targeted (root system 1, root system 2) as shown in Figure 1. The trees grew in a coastal forest stretching 8 km to the west end of the Atsumi Peninsula, Tahara City, Aichi, Tokai District, Japan [32–35], and were replanted after heavy damage by Typhoon Vera in 1959 [32]. The soil was sandy in nature [36]. From 1981 to 2010, the mean annual temperature was 16.0 °C, and the mean annual precipitation was 1603 mm [37].

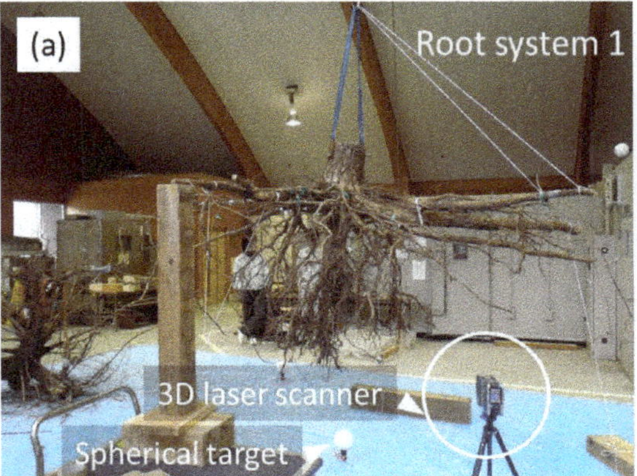

Figure 1. Measurement of (**a**) Root system 1, (**b**) Root system 2 by 3D laser scanner.

After measuring the tree height and DBH, we measured the critical turning moment in tree-pulling experiments using the method described by Todo et al. [35] and cut the trees at the ground level. The tree-pulling experiment was performed as follows: a polyester belt sling (safe working load: 6.3 t) was attached to the tree, connected to a 10 mm diameter wire rope 1 m above the ground, and this was pulled, parallel to the ground, by an excavator. A load cell (maximum load 50 kN, LT-50KNG56 NIKKEI Electronic Instruments Co. Ltd., Tokyo, Japan) was connected between the belt sling and wire rope, and the load data were recorded at intervals of 0.1 s through a bridge unit (DBU-120A, KYOWA Electronic Instruments Co. Ltd., Tokyo, Japan). Measurements of the loads for the tree-pulling experiments began before the excavator commenced pulling the wire rope and stopped when the load began to decrease after reaching the maximum. In this study, the test trees were not uprooted because the maximum critical turning moment has already reached before [35] and the root system could keep the positions in soils.

The trees were aged from the rings in the stump, and the whole root system was dug out using an air spade [33,34] in January 2014. The two trees were 45 and 50 years old, the DBHs were 19.5 and 18.5 cm, and the heights were 10.8 m and 11.4 m (Table 1), respectively. The maximum root depth was 126 cm in root system 1 and 106 cm in root system 2. Both trees had tap root systems (Figure 1). On the other hand, the critical turning moment in root system 2 was approximately 1.3 times that in root system 1, probably owing to different RSAs (Table 1). The excavated root systems were placed inside a room so as to preserve the 3D RSA.

Table 1. Properties of *Pinus thunbergii* trees.

Properties	Root System 1	Root System 2
Age (y)	45	50
Height (m)	10.8	11.4
Stem diameter at breast height (cm)	19.5	18.5
Critical turning moment (kN·m)	25.5	33.6
Maximum depth of root (cm)	126	106

2.2. 3D Laser Scanner Measurement

Point cloud data on the surface of the root system (Figures 2–4) were collected using 3D laser scanners (FARO Focus3D S120, FARO Technologies Inc., Lake Mary, FL, USA) (Figure 1). The two root systems were suspended while facing up in a room and supported to maintain the original RSA. The suspended root system was positioned by keeping the ground surface portion of the stump horizontal and aligning it with the orientation of the site. The two root systems were individually scanned from eight positions using a 3D laser scanner (FARO S120) within 2 m from the center of the collar. Scanning was performed to cover the entire root system, and three spherical targets were identified in each scan. The scanner had a measurement range of 0.6 m to 120 m with a vertical field of view of 300° and a horizontal field of view of 360°. The angular resolution was 0.009° for both horizontal and vertical angles. The scanner output was a 905 nm laser beam with a 3 mm circular diameter and 0.19 mrad divergence. The scanner had a stated error of 2 mm between the scanner position and the shooting object and a standard deviation in the optimum plane scan of $\leq$2 mm [38]. The point cloud data taken from each position were integrated into one 3D point cloud data by matching the reflected spherical targets. FARO Scene (FARO Technologies, Inc., version 5.1.6) was used to integrate the point cloud data.

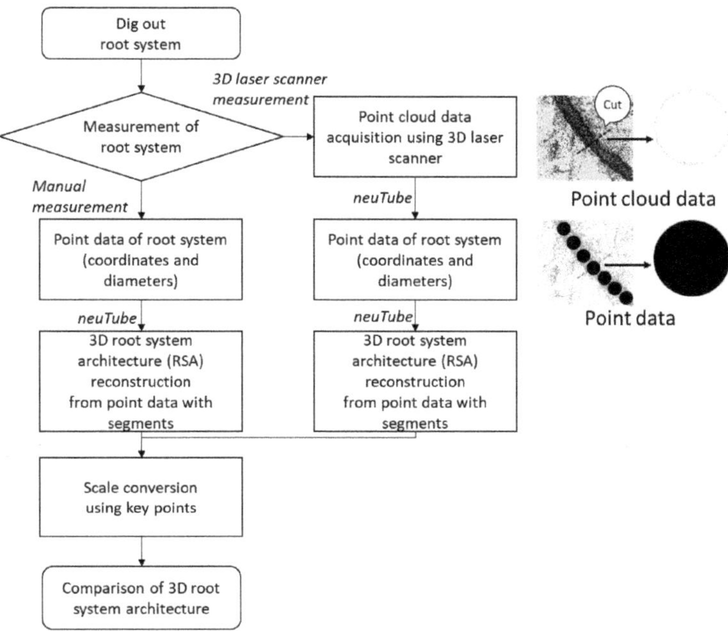

Figure 2. Acquisition of point cloud data and root point data and reconstruction of root system architecture (RSA).

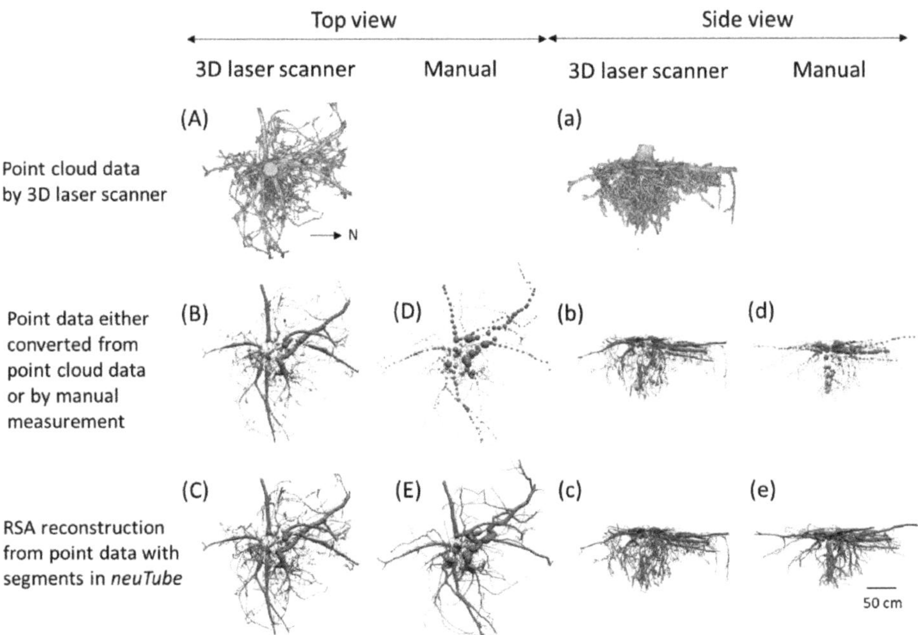

Figure 3. (**A**–**E**) Top and (**a**–**e**) side views of Root system 1 drawn from (**A**,**a**) point cloud data collected by 3D laser scanner; root point data; (**B**,**b**) converted from point cloud data; (**D**,**d**) gathered by hand; and (**C**,**E**,**c**,**e**) RSA reconstructed by neuTube software.

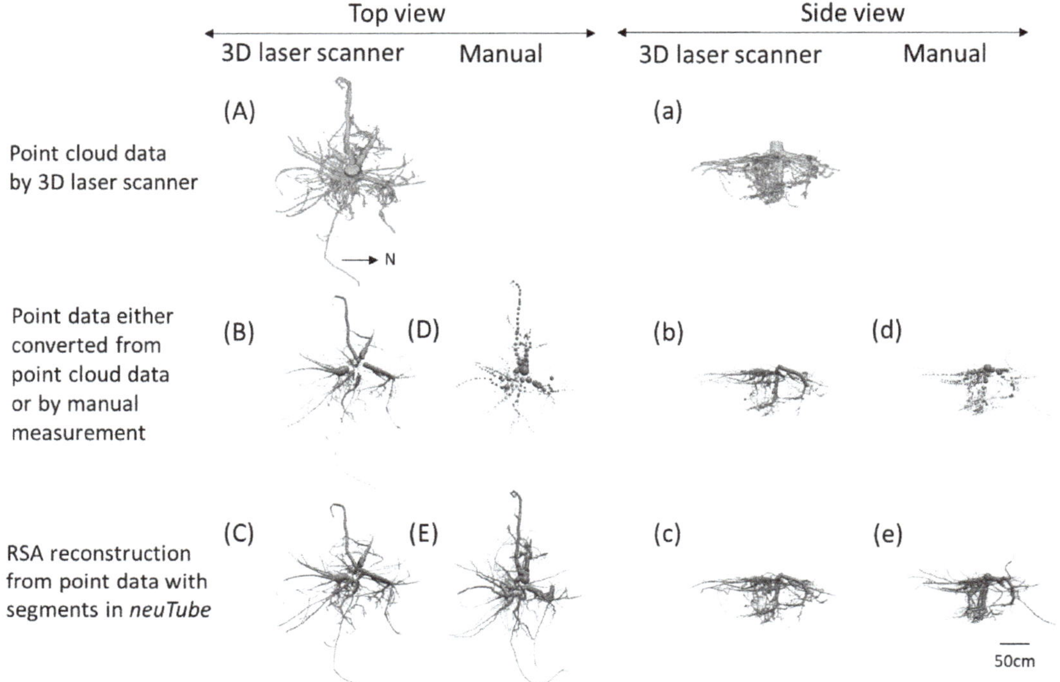

Figure 4. (**A–E**) Top and (**a–e**) side views of Root system 2 drawn from (**A**,**a**) point cloud data collected by 3D laser scanner root point data; (**B**,**b**) converted from point cloud data; (**D**,**d**) gathered by hand; and (**C**,**E**,**c**,**e**) RSA reconstructed by neuTube software.

We used free software designed for digital reconstruction of neuron morphologies, neuTube [39], to generate structural models of the roots from the image data. One method used to measure the morphology of neurons with complex branching structures is to obtain a series of images of fluorescently stained neurons while changing the focal plane (confocal microscopic images) and then reconstruct the 3D structure. As neuTube was developed to analyze the branching structure of neurons from confocal image data, we thought it may be useful in analyzing tree root system data.

First, we extracted the point cloud data from only the root surfaces by deleting other noise data using FARO Scene's scatter filter and then manually eliminating the remaining noise. Second, we stratified the point cloud data by depth and created a series of 2D cloud image data points every 1 cm. As the point cloud data describe only the root surfaces, the outer periphery of the root CSA appears in each image. Third, the area enclosed by each circle along the long axis of each root segment was filled manually and visually using ImageJ software. Even in an unclosed circle such as straight horizontal roots, we could identify it and fill between unclosed parallel lines manually and visually. The series of sliced images were loaded into neuTube [39], and the automatic tracing function tracked each circle in the long axis direction, outputting data related to the diameter at the branch points and connections between roots, that is, the information of root point data. The number of root point data converted from the point cloud data in root systems 1 and 2 was 5885 and 2301, respectively. All these steps, including those involving neuTube, can be performed on a regular PC. The cross-section of each segment was assumed to be a circle in this software. In this study, we did not collect point data of the stumps from the point cloud data because we focused on the possibility of reconstruction of the root system, particularly the traits of the main roots and root segments.

2.3. Manual Measurement

The root point data were also acquired through manual measurements [13]. Each root system was fixed upside down over a sheet marked with a 100 mm × 100 mm grid that served as a guide for the x and y coordinates (Figure S1). The horizontal level of the stump was maintained at the ground surface to adjust the root position, the y-axis direction was fixed to the north, and the flexible roots were fixed in stands while comparing the photos taken in the field to those during the 3D scan. Roots whose z coordinates were displaced by being upside down were adjusted on a stand. After cutting off thin roots with a diameter of <5 mm, we measured the 3D coordinates (x, y, z) and the diameter of roots at root end points, branch points, and where the grid lines intersected with the roots (Figure S1). Each root point was assigned an ID number, and the connections between the points were recorded. The horizontal coordinates (x, y) were measured with a steel tape with reference to the grid lines, and the x and y coordinates were determined using the center of the root collar as the origin. The z coordinates were determined by measuring the distance from the measurement point on the root system to the ground along a plumb line, using the ground surface as the origin. The root diameter was measured vertically and horizontally using a digital caliper and averaged. The number of measured root data points was 1791 for root system 1 and 1990 for root system 2.

2.4. Reconstruction of RSA from Root Point Data

The root point data set, either calculated from the point cloud data or measured manually, was used to reconstruct the RSA (Figures 3 and 4). Pairs of adjacent points were connected to the approximate frustum of the cones. neuTube was used to analyze the connecting root point data and to display the roots. Since neuTube can import morphological models based on the SWC file format [39], we saved both scanner and manual measurement data in SWC format. The SWC format can allow the creation of MTG coding files, which are often used to study the relationship between RSA and anchorage [40].

2.5. Calibration of RSA Data

RSA data were acquired as pixels. We selected 10 identifiable key points (root end points and inflection points) in each root system (Figure S2), compared their xyz coordinates and diameters between the 3D laser scanner data (in pixels) and the manual measurements (in mm), and derived a formula to convert scanner data to mm (Table S1).

2.6. Comparison of RSA Data Obtained by Different Methods

To evaluate the differences in RSA data at the root point level between the methods, we compared the xyz coordinates and root diameters at another 10 points in each root system (Figure 5). The differences (Di) in distances from each coordinate obtained by the manual measurement to that of the 3D laser scanner were calculated at the 10 root points in each root system. The accuracy (Ac) of root diameter at the 10 root points in each root system was defined as the percentage of the diameter estimated by the 3D laser scanner to that of the manual measurement according to Lau et al. [27]. To evaluate the differences across all RSA data in each root system, we compared the top and side views of the point cloud data with the RSA data reconstructed from the laser scanner data, and the latter with the manual measurements.

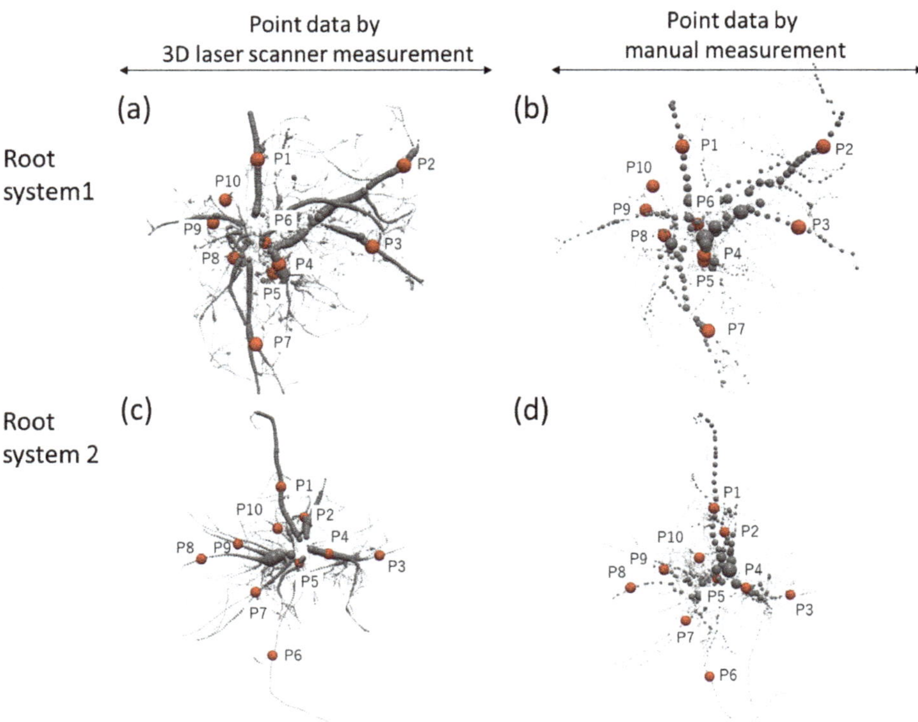

Figure 5. Positions of 10 points (P1–P10) set to evaluate the accuracy of the *xyz* coordinates and root diameters in root point data collected in (**a**,**b**) Root system 1 and (**c**,**d**) Root system 2 by (**a**,**c**) 3D laser scanner and (**b**,**d**) hand.

We compared the change in root diameter (degree of root taper) in the direction of root extension between the measurement methods. We selected 10 main root segments in each reconstructed RSA (Figure 6) and compared the diameters between the methods. The accuracy (Ac) of the diameter in each root segment was defined as shown above. We also compared the differences between the methods in total root CSA in the vertical direction at a given horizontal distance from the collar center in the reconstructed RSAs. To approximate the root–soil plate as an elliptical pillar or a cylinder, it was necessary to measure the number, diameter, and CSA of the roots on the side of the cylinder (the root–soil plate). The total root CSA of the reconstructed RSA was calculated from the cross-sectional diameter at concentric columnar surfaces every 100 mm (between 300 mm and 1800 mm) from the collar center. The accuracy (Ac) of CSA was defined as the percentage of CSA by the 3D laser scanner to that of the manual measurement in each root system.

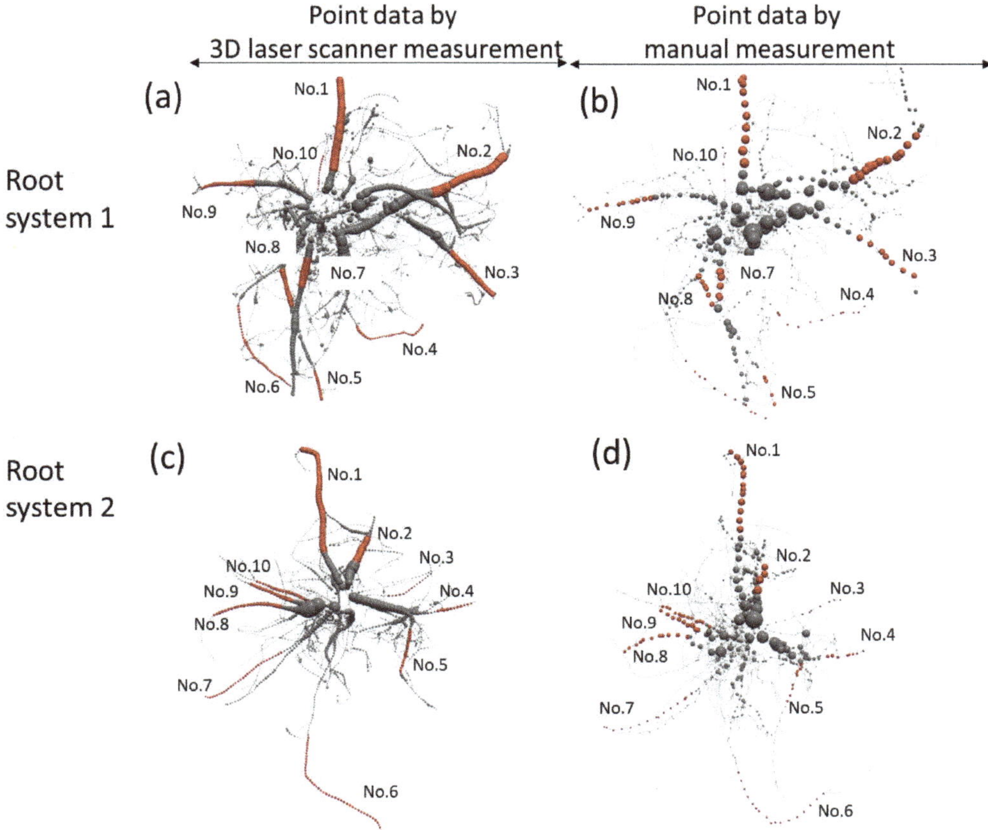

Figure 6. Ten main root segments in each root system selected to compare tapering between (**a**,**b**) in Root system 1; and (**c**,**d**) in Root system 2 measured by (**a**,**c**) 3D scanner and (**b**,**d**) hand.

2.7. Statistical Analysis

Normal distribution of *xyz* coordinates and diameters of 10 selected root point data, the diameters of 10 root segments along the long axis data, and total root CSA data in each root system was confirmed by the Kolmogorov–Smirnov test. We conducted a paired *t*-test to compare the differences in the diameters of the 10 root segments along the long axis data obtained by the manual measurement with those measurements obtained by the 3D laser scanner. We produced correlations between the manual and 3D laser scanner measurements in *xyz* coordinates and the diameter of 10 selected root point data, and total CSA data. Because of the small sample size for each trait, we calculated bootstrapped estimates of the correlation coefficient for 1000 iterations and showed the values in the results. Statistical analyses were conducted using the R software version 4.0.2 [41].

3. Results

3.1. Reconstruction of RSA in Manual and 3D Laser Scanner Measurements

We successfully reconstructed the RSAs of the two *P. thunbergii* trees using neuTube software (Figures 3 and 4). Visually, the views from the top (Figure 3B,C and Figure 4B,C) and side (Figure 3b,c and Figure 4b,c) of the 3D laser scanner RSAs, which were very similar to those of the point clouds (Figure 3A,a and Figure 4A,a; Figures S3a,b and S4a,b), and the main root segments were reproduced. However, the reconstructed RSAs were more sparse and were simpler than point clouds.

The manual data RSAs (Figure 3D,E,d,e and Figure 4D,E,d,e) were reconstructed to a similar extent as in the 3D laser scanner RSAs (Figure 3B,C,b,c and Figure 4B,C,b,c) based on visual comparison. Similarly, the main root segments were well reproduced. However, the manual data RSAs were simpler and sparser than the surface-shaped 3D laser scanner RSAs (Figure 3A,a and Figure 4A,a).

3.2. Comparison of Root Point Data between Manual and 3D Laser Scanner Measurements

The positions and root diameters at the 10 points in the root point data were positively correlated with the 3D laser scanner and the manual measurements in root systems 1 and 2 ($p < 0.01$; Figure 7). The differences (Di) in the distance of x, y, and z coordinates were −1.4, −7.7, and −19.0 mm in root system 1 and −45.7, 20.5, and −7.7 mm in root system 2, respectively (Figure 7a–c). Roots with a diameter of approximately 7 to 50 mm were measured accurately in the root point data converted from the 3D laser scanner data (Figure 7d). The accuracies (Ac) of the diameters were 97.7% and 86.6% in root systems 1 and 2, respectively.

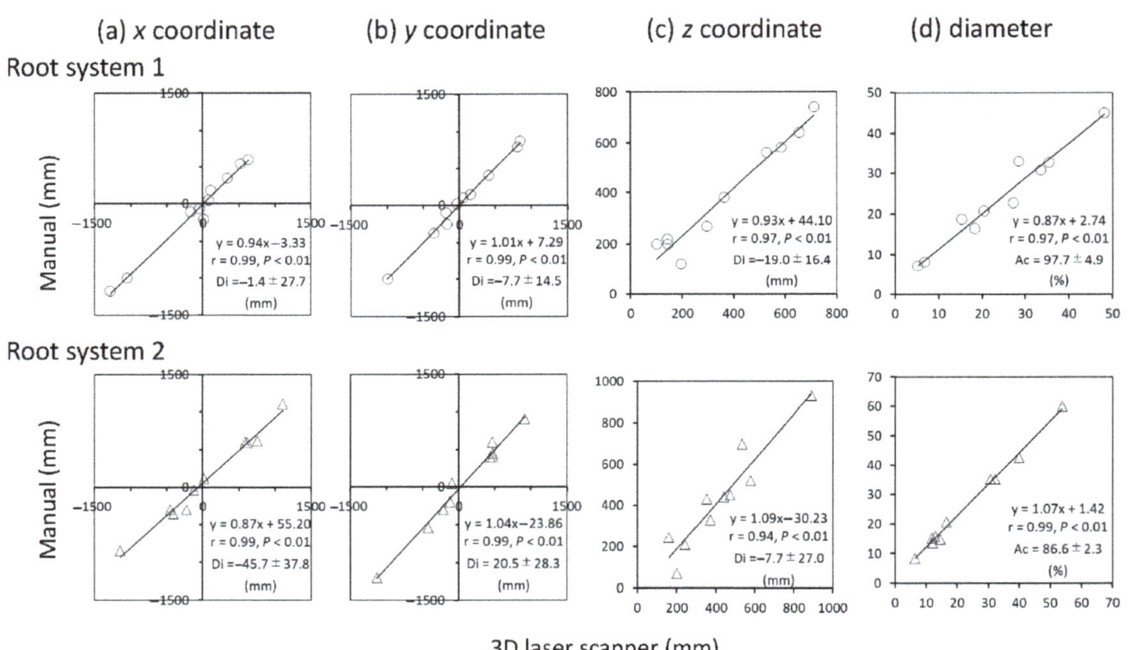

Figure 7. Relationships of (**a**) x, (**b**) y, (**c**) z coordinates, and (**d**) diameter of 10 selected root points between 3D laser scanner and manual measurements. The correlation coefficient (r) was calculated as the mean value estimated by 1000 bootstraps. The differences (Di, mean ± standard error) in distances from each coordinate obtained by the manual measurement to that of the 3D laser scanner were calculated at the 10 root points in each root system. The accuracy (Ac, mean ± standard error) of root diameter at 10 root points in each root system was defined as the percentage of the diameter estimated by the 3D laser scanner to that of the manual measurement according to Lau et al. [27].

3.3. Differences in Taper and CSA of Roots between Manual and 3D Laser Scanner Measurements

In a visual comparison, the taper trends of the 20 root segments in both root systems were similar between the 3D laser scanner and manual measurements, with variation (Figure 8). The mean difference in root diameter between the two measurement methods was 4.0 mm, and the maximum was 16.4 mm. The mean diameters of 14 of the 20 root segments estimated by manual measurement were not significantly different from the 3D laser scanner ($p > 0.05$; Figure 8). Although the diameters of the remaining six root segments were significantly different between the measurements ($p < 0.05$; Figure 8), the differences were minimal (−7.44 and 7.94 mm). The average accuracy (Ac) of the 3D laser scanner measured diameter compared to the manually measured diameter of the root segment ranged between 80.1% to 117.0% for root system 1 and from 83.0% to 145.1% for root system 2, respectively (Figure 8).

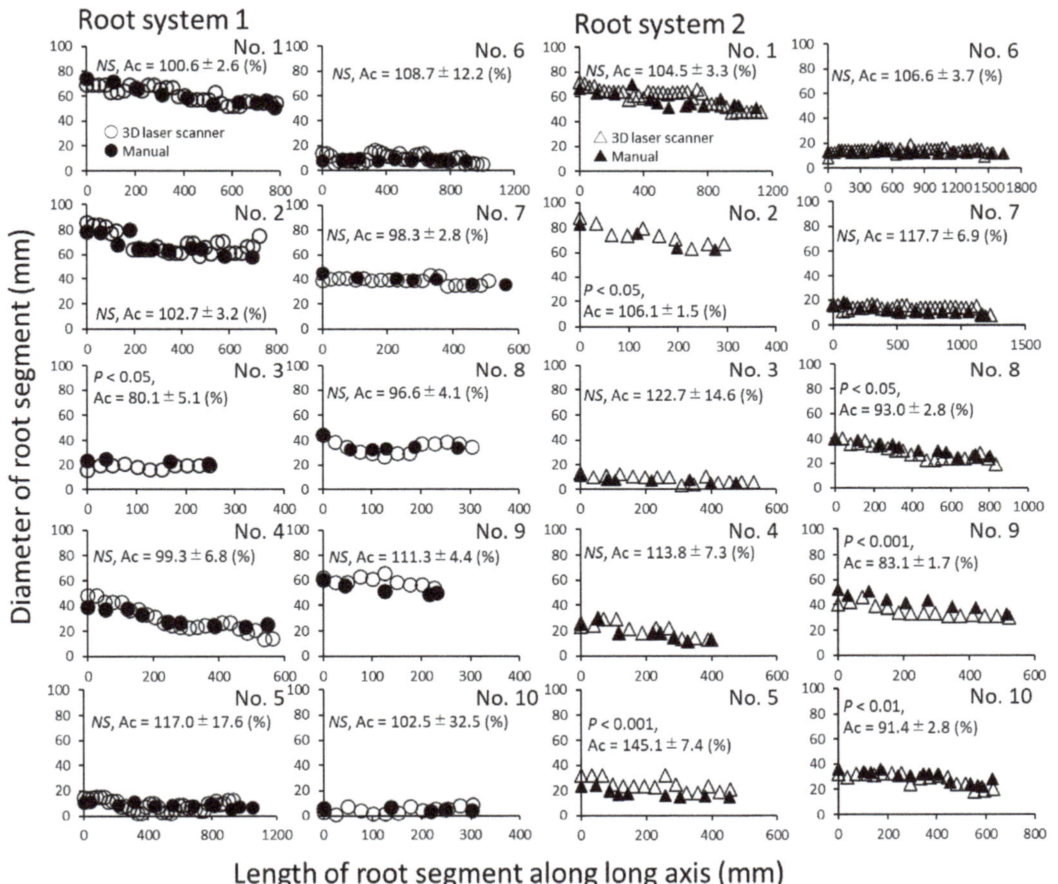

Figure 8. Changes in the diameters of 10 root segments along the long axis in each root system by 3D laser scanner and manual measurements. A paired *t*-test was used to compare the differences in the diameters of 10 root segments along the long axis data obtained by the manual measurement with those obtained by the 3D laser scanner. NS: not significant ($p > 0.05$). The correlation coefficient (r) was calculated as the mean value estimated by 1000 bootstraps. The accuracy (Ac, mean ± standard error) of the diameter at 10 root segments in each root system was defined as the percentage of the diameter estimated by the 3D laser scanner to that of the manual measurement according to Lau et al. [27].

The total CSAs tended to decrease with distance from the collar center in both methods (Figure 9a,b), notably so in root system 1 at <500 mm. The correlations of the total CSAs between the measurements were consistently high in each root system (root system 1, r = 0.90; root system 2, r = 0.93; Figure 9a,b). The accuracy (Ac) of CSA was 103.0% in root system 1 and 95.7% in root system 2. There was a positive correlation between the methods, with a slope of 0.82 (r = 0.88), as seen in Figure 9c. This indicated that the total CSAs estimated by the 3D laser scanner were smaller than those estimated from manual measurements.

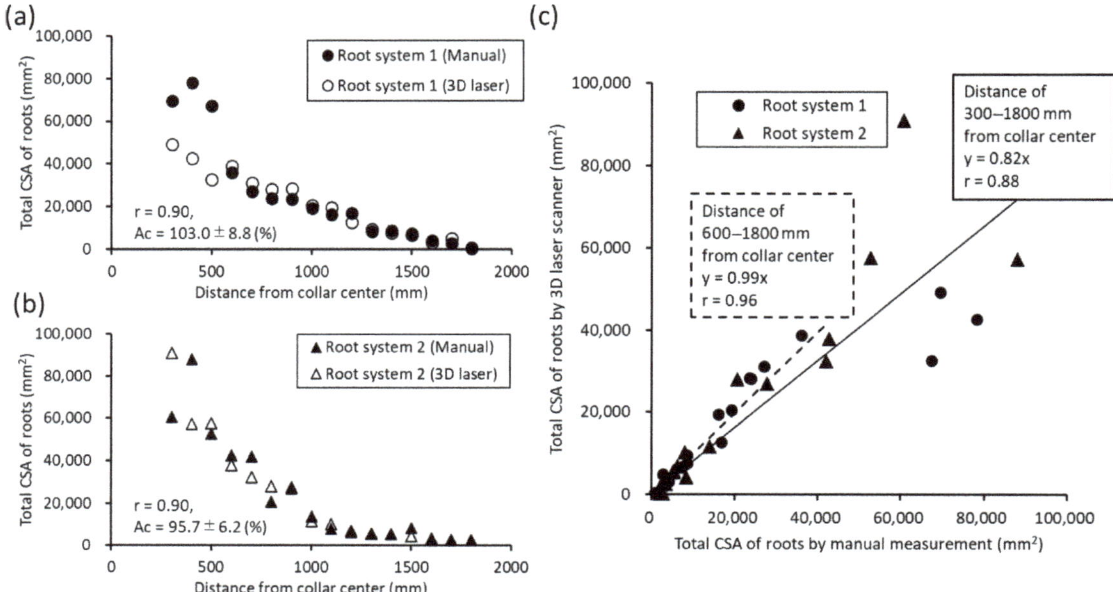

Figure 9. Total cross-sectional area (CSA) of (**a**) Root system 1 and (**b**) Root system 2 with distance from the collar center by 3D laser scanner and manual measurements. (**c**) Relationship of the total CSA of both root systems between 3D laser scanner and manual measurements. Paired t-test was used to compare the differences in the CSA at the distances from the collar center obtained by the manual measurement with those by the 3D laser scanner measurement. NS: not significant ($p > 0.05$). The correlation coefficient (r) was calculated as the mean value estimated by 1000 bootstraps. The accuracy (Ac, mean ± standard error) of the CSA at the distances from the collar center in each root system was defined as the percentage of the CSA estimated by the 3D laser scanner to that of the manual measurement according to Lau et al. [27].

4. Discussion

4.1. Significant Steps in Reconstructing RSA from Point Cloud Data

We proposed an innovative method for reconstructing RSA and calculating traits such as taper and CSA of roots from root point data converted from point cloud data collected by a 3D laser scanner. This is the first application of the neuTube software, which was developed to reconstruct the neuronal structure [39], to the RSA of forest trees. The accuracy was very similar to that of the manually collected root point data. Although previous studies have scanned excavated root systems using a 3D laser scanner [28,31,42], the accuracy of the resultant root system models have not been evaluated against intensive manual measurements [31]. Moreover, one drawback of measuring RSA using a 3D laser scanner is that it is obtained as point cloud data [12].

Lau et al. [27] reported an absolute error of 5.14 cm in the diameter class of 10–20 cm of tree branches and −3.46 cm in the 20–60 cm class when the data calculated by the QSM model using the point cloud data by the 3D laser scanner were compared with that of the manual measurement data. The accuracy (Ac) of the root diameter estimated by our proposed methods using the point cloud data of the 3D laser scanner measurement to the manually measured root diameter was 98% for root system 1 and 87% for root system 2 (Figure 7c). The difference in root diameter between the methods might have occurred due to the difference in the degree of dryness of the roots at the time of measurement.

The maximum difference in distance (Di) between the coordinates of the roots was less than 46 mm, and the minimum was 1.4 mm (Figure 7a–c). These results indicated that our proposed method, without any specific parameters, can be reproduced with high accuracy using a 3D laser scanner. Our proposed method can easily convert point cloud data to root point data and RSA data and calculate RSA traits. Such analysis is important for examining the resistance of trees to uprooting and landslides [2,11,43]. Information on root traits can only be obtained through the direct measurement of the root system. However, if the root point data or point cloud data are converted to RSA data, further details can be obtained from the stored data if required.

Another advantage of 3D laser scanner measurement is that it requires less labor and time to collect data where root systems were set either outside or inside the laboratory compared to manual and semi-automatic methods [12]. In this study, collecting *xy* coordinates by hand at intervals of 10 cm and at the branches and tips of roots took three people and five days per root system. In contrast, 3D laser scanners set at eight points scanned the root system in 60 min. By collecting a large number of data points, we can reconstruct subtle root irregularities and changes in root diameter that manual measurement would not reveal.

On the other hand, for the data treatment after the acquisition, 3D laser scanner measurement took one person 15 h per root system to eliminate noise from the point cloud data in 2014. However, the recent software for point cloud analysis has improved the performance and reduced the time required for noise elimination. Our process of converting the point cloud data of the root system into root point data was partly manual using ImageJ. It took about one person-hour to complete the conversion of the whole root system, and more than 2000 root point data were made. Thus, it took approximately 2.0 person-days to obtain the root point data from the point cloud data using a 3D laser scanner. The semi-automated method using 3D digitizing also took 2.5 person-days to obtain the whole root system in a 14-year *P. pinaster* with a DBH of 17.0–17.4 cm and the height of 9.7–10.9 m [6]. However, the method proposed in our study did not include the time for classifying segments in several root compartments, such as the ZRT. On the other hand, semi-automatic measurements provide the direction of ovality, which is required to compute flexural stiffness in a given direction of load [6]. Therefore, the 3D laser scanner method is inferior to the semi-automated method using 3D digitizing. If the time required for noise processing of point cloud data can be reduced and the work in ImageJ can be automated, further time reduction can be expected in the proposed method.

4.2. Other Issues to Consider in the Use of 3D Laser Scanning

There are some issues in the acquisition and analysis of point cloud data using 3D scanner measurements. First, when root systems are scanned, the positions of the excavated roots should be the same as that of the soil as much as possible, especially in the vertical direction. In our results, root system 1 and the *z* coordinates, which indicate root depth, differed slightly between methods with the *x* and *y* coordinates (Figure 7). This may be because we scanned the root systems upright but measured them while upside down. Thin horizontal roots can be bolted to a plate [14] or tied to a stage or pole [28], but plenty of space is needed.

Second, the CSA of roots close to the collar, especially within 500 mm, differed between methods (Figure 9), perhaps because the laser beam did not reach the whole root surface, owing to their proximity to each other and the collar. In fact, the slope of the relationship of CSA between the two methods was 0.99 at 600–1800 mm from the collar center ($r = 0.96$; Figure 9c), but was only 0.82 at 300–1800 mm, indicating that the CSA of roots near collars was larger by manual measurement than by 3D laser scanner measurement. The error can be reduced by: (i) taking a more detailed scan or using pointer lamps [28], where errors are likely to occur; and (ii) finding a location where the root diameter changes notably and then using a correction.

Finally, a priority task is to tackle the problem of cross-sectional ovality, model the stump, and convert it into a file for RSA analysis (e.g., MTG file) to obtain a complete pipeline of measurements. neuTube assumes a circular cross-sectional shape; thus, each root segment is a frustum of a cone. This assumption is unsuitable for representing the heterogeneous shapes of root cross-sections, particularly for T- and I-shapes formed under intense stress [44]. Therefore, it is necessary to adapt this method to tree species with non-cylindrical roots. Furthermore, in this study, we did not collect point data from the stump, which is an important part of considering the whole root system and plays a specific role in anchorage [45].

Resolving these issues would allow for more accurate RSA data acquisition from 3D laser scanner measurements using our proposed method. As other reconstruction methods for modeling or branch architecture exist, comparison with these methods would be ideal to elucidate the advantages and disadvantages depending on the objective of the reconstruction [46].

The limitation of this study is that it focused on only one tree species of two *Pinus thunbergii* trees that had typical tap root systems. To establish it as a practical method, the estimated root traits in several tree species with different root system types and the advantages and disadvantages of the method during the measurement should be compared with the most common semi-automatic measurement methods [7].

4.3. Future Aspects for RSA Measurement

3D measurement requires digging out the root system, so time-series data cannot be collected. Because root point data obtained by non-destructive methods, such as ground-penetrating radars [47], do not indicate connections, it is difficult to convert them into RSA data without a connection algorithm [48]. Recently, we devised a connection algorithm for point data of *Cryptomeria japonica* roots [13]. Such methods can be used to process root point data without connection information, but it is still necessary to excavate root systems and measure root positions, diameters, and connections manually before devising an algorithm [13]. Our results confirm that the root point data converted from the point cloud data and the reconstructed RSA are very similar to the manual data. Therefore, using 3D laser scanner data, we can develop connection algorithms for the root systems of other tree species.

5. Conclusions

We proposed a new method using the neuTube software to convert point cloud data to root point data to reconstruct the RSA of *P. thunbergii*. The RSAs proved to be accurate. The root point data were connected with frustum of the cones, and thus, the reconstructed RSAs were relatively simple but included the main coarse root segments. Although a 3D laser scanner can collect detailed data in a short time, it is expensive, the root system must be brought inside for measurement, and specific technology and time are required for imaging. In contrast, manual measurement requires intensive labor and time but does not require special machinery or technology. Semi-automatic measurements can address these challenges [11,12,20]. Essentially, RSA data can be obtained efficiently by 3D laser scanner analysis to evaluate the root anchorage and the slope stability of forest trees, provided that we could succeed in modeling both the stump and ovality of root sections.

Supplementary Materials: The following are available online at https://www.mdpi.com/article/10.3390/f12081117/s1, Table S1: Conversions of root point data from 3D laser scanner to scale based on the relationships in the coordinates or diameter of roots between manual measurement data (xm, ym, zm, dm) and 3D laser scanner measurement data (x3D, y3D, z3D, d3D). Figure S1: How we measured the root system by hand against a 100 mm × 100 mm grid. This photo shows *Cryptomeria japonica*. Figure S2: Positions of 10 points (K1–K10) set to relate the locations of roots to the actual scale. Figure S3: (a) Top and (b) side views of root system 1 drawn from point cloud data collected by the 3D laser scanner. These are the same as in Figure 3 (A,a) but enlarged. Figure S4: (a) Top and (b) side views of root system 2 drawn from point cloud data collected by the 3D laser scanner. These are the same as in Figure 4 (A,a) but enlarged.

Author Contributions: C.T.: planned and performed the experiments, analyzed the data, wrote the paper, and prepared it for submission; H.I.: planned and performed the experiments, analyzed the data, and wrote the paper; K.Y., T.T., and M.D.: performed the experiments and analyzed the data; M.O. and T.K.: analyzed the data; Y.H.: planned and performed the experiments, analyzed the data, and wrote and edited the paper. All authors have read and agreed to the published version of the manuscript.

Funding: This study was partly supported by JSPS KAKENHI grants (JP25252027 and 2003028).

Data Availability Statement: Data available in a publicly accessible repository that does not issue DOIs. Publicly available datasets were analyzed in this study. These data can be found at https://github.com/HidetoshiIkeno/3D-RSA (accessed on 14 July 2021).

Acknowledgments: We appreciate the editor and reviewers for providing constructive comments and suggestions for the enhancement of our manuscript. We thank Y. Kimura, Y. Nakagawa, T. Akamatsu, A. Takamatsu, and M. Yoshida (University of Hyogo) for collecting the data manually; H. Ogata, H. Nakamura (woodinfo Co., Ltd.) for the 3D scanning, and J. Tanaka (Japan Conservation Engineers & Co., Ltd.) and Aichi Prefecture for allowing us to conduct the experiment at the study site.

Conflicts of Interest: The authors declare that they have no conflict of interest.

References

1. Jackson, R.B.; Pockman, W.T.; Hoffmann, W.A. The Structure and Function of Root Systems. In *Handbook of Functional Plant Ecology*; Pugnaire, F.I., Valladares, F., Eds.; Marcel Dekker, Inc.: New York, NY, USA, 1999; pp. 196–220.
2. Schwarz, M.; Lehmann, P.; Or, D. Quantifying Lateral Root Reinforcement in Steep Slopes—From a Bundle of Roots to Tree Stands. *Earth Surf. Process Landf.* **2010**, *35*, 354–367. [CrossRef]
3. Yamase, K.; Todo, C.; Torii, N.; Tanikawa, T.; Yamamoto, T.; Ikeno, H.; Ohashi, M.; Dannoura, M.; Hirano, Y. Dynamics of Soil Reinforcement by Roots in a Regenerating Coppice Stand of *Quercus serrata* and Effects on Slope Stability. *Ecol. Eng.* **2021**, *162*, 106169. [CrossRef]
4. Ennos, A.R. The Mechanics of Root Anchorage. *Adv. Bot. Res.* **2000**, *33*, 133–157.
5. Ghestem, M.; Veylon, G.; Bernard, A.; Vanel, Q.; Stokes, A. Influence of Plant Root System Morphology and Architectural Traits on Soil Shear Resistance. *Plant Soil* **2014**, *377*, 43–61. [CrossRef]
6. Dorval, A.D.; Meredieu, C.; Danjon, F. Anchorage Failure of Young Trees in Sandy Soils Is Prevented by a Rigid Central Part of the Root System with Various Designs. *Ann. Bot.* **2016**, *18*, 747–762. [CrossRef] [PubMed]
7. Danjon, F.; Stokes, A.; Bakker, M.R. Root Systems of Woody Plants. In *Plant Roots; The Hidden Half*; CRC Press: Boca Raton, FL, USA, 2013; p. 848.
8. Yang, M.; Défossez, P.; Danjon, F.; Fourcaud, T. Analyzing Key Factors of Roots and Soil Contributing to Tree Anchorage of *Pinus* species. *Trees* **2018**, *32*, 703–712. [CrossRef]
9. Liang, T.; Knappett, J.A.; Bengough, A.G.; Ke, Y.X. Small-Scale Modelling of Plant Root Systems Using 3D Printing, with Applications to Investigate the Role of Vegetation on Earthquake-Induced Landslides. *Landslides* **2017**, *14*, 1747–1765. [CrossRef]
10. Yamase, K.; Tanikawa, T.; Dannoura, M.; Todo, C.; Yamamoto, T.; Ikeno, H.; Ohashi, M.; Aono, K.; Doi, R.; Hirano, Y. Estimating Slope Stability by Lateral Root Reinforcement in Thinned and Unthinned Stands of *Cryptomeria japonica* Using Ground-Penetrating Radar. *Catena* **2019**, *183*, 104227. [CrossRef]
11. Danjon, F.; Fourcaud, T.; Bert, D. Root Architecture and Wind-Firmness of Mature *Pinus pinaster*. *New Phytol.* **2005**, *168*, 387–400. [CrossRef]
12. Danjon, F.; Reubens, B. Assessing and Analyzing 3D Architecture of Woody Root Systems, a Review of Methods and Applications in Tree and Soil Stability, Resource Acquisition and Allocation. *Plant Soil* **2008**, *303*. [CrossRef]
13. Ohashi, M.; Ikeno, H.; Sekihara, K.; Tanikawa, T.; Dannoura, M.; Yamase, K.; Todo, C.; Tomita, T.; Hirano, Y. Reconstruction of Root Systems in *Cryptomeria japonica* Using Root Point Coordinates and Diameters. *Planta* **2019**, *249*, 445–455. [CrossRef]
14. Henderson, R.; Ford, E.D.; Renshaw, E.; Deans, J.D. Morphology of the Structural Root System of Sitka spruce 1. Analysis and Quantitative Description. *Forestry* **1983**, *56*, 121–135.
15. Mulatya, J.M.; Wilson, J.; Ong, C.K.; Deans, J.D.; Sprent, J.I. Sprent. *Agrofor. Syst.* **2002**, *56*, 65–72. [CrossRef]
16. Oppelt, A.L.; Kurth, W.; Dzierzon, H.; Jentschke, G.; Godbold, D.L. Structure and Fractal Dimensions of Root Systems of Four Co-Occurring Fruit Tree Species from Botswana. *Ann. For. Sci.* **2000**, *57*, 463–475. [CrossRef]
17. Godin, C.; Caraglio, Y.; Costes, E. Exploring Plant Topological Structure with the AMAPmod Software: An Outline. *Silva Fenn.* **1997**, *31*, 357–368. [CrossRef]

18. Danjon, F.; Sinoquet, H.; Godin, C.; Colin, F.; Drexhage, M. Characterisation of Structural Tree Root Architecture Using 3D Digitising and AMAPmod Software. *Plant Soil* **1999**, *211*, 241–258. [CrossRef]
19. Sorgonà, A.; Proto, A.R.; Abenavoli, L.M.; Di Iorio, A. Spatial Distribution of Coarse Root Biomass and Carbon in a High-Density Olive Orchard: Effects of Mechanical Harvesting Methods. *Trees* **2018**, *32*, 919–931. [CrossRef]
20. Cast, C.S.; Meredieu, C.; Défossez, P.; Pagès, L.; Danjon, F. Modelling Root System Development for Anchorage of Forest Trees up to the Mature Stage, Including Acclimation to Soil Constraints: The Case of *Pinus pinaster*. *Plant Soil* **2019**, *439*, 405–430. [CrossRef]
21. Lontoc-Roy, M.; Dutilleul, P.; Prasher, S.O.; Han, L.; Smith, D.L. Computed Tomography Scanning for Three-Dimensional Imaging and Complexity Analysis of Developing Root Systems. *Can. J. Bot.* **2005**, *83*, 1434–1442. [CrossRef]
22. Calders, K.; Newnham, G.; Burt, A.; Murphy, S.; Raumonen, P.; Herold, M.; Culvenor, D.; Avitabile, V.; Disney, M.; Armston, J.; et al. Nondestructive Estimates of Above-Ground Biomass Using Terrestrial Laser Scanning. *Methods Ecol. Evol.* **2015**, *6*, 198–208. [CrossRef]
23. Dassot, M.; Constant, T.; Fournier, M. The Use of Terrestrial LiDAR Technology in Forest Science: Application Fields, Benefits and Challenges. *Ann. For. Sci.* **2011**, *68*, 959–974. [CrossRef]
24. Raumonen, P.; Kaasalainen, M.; Åkerblom, M.; Kaasalainen, S.; Kaartinen, H.; Vastaranta, M.; Holopainen, M.; Disney, M.; Lewis, P. Fast Automatic Precision Tree Models from Terrestrial Laser Scanner Data. *Remote Sens.* **2013**, *5*, 491–520. [CrossRef]
25. Seidel, D.; Fleck, S.; Leuschner, C.; Hammett, T. Review of Ground-based Methods to Measure the Distribution of Biomass in Forest Canopies. *Ann. For. Sci.* **2011**, *68*, 225–244. [CrossRef]
26. Srinivasan, S.; Popescu, S.; Eriksson, M.; Sheridan, R.; Ku, N. Terrestrial Laser Scanning as an Effective Tool to Retrieve Tree Level Height, Crown Width, and Stem Diameter. *Remote Sens.* **2015**, *7*, 1877–1896. [CrossRef]
27. Lau, A.; Bentley, L.P.; Martius, C.; Shenkin, A.; Bartholomeus, H.; Raumonen, P.; Malhi, Y.; Jackson, T.; Herold, M. Quantifying Branch Architecture of Tropical Trees Using Terrestrial LiDAR and 3D Modelling. *Trees* **2018**, *32*, 1219–1231. [CrossRef]
28. Gärtner, H.; Wagner, B.; Heinrich, I.; Denier, C. 3D-Laser Scanning: A New Method to Analyze Coarse Tree Root Systems. *For. Snow Landsc. Res.* **2009**, *82*, 95–106.
29. Wagner, B.; Gärtner, H.; Ingensand, H.; Santini, S. Incorporating 2D tree-ring data in 3D laser scans of coarse-root systems. *Plant Soil* **2010**, *334*, 175–187. [CrossRef]
30. Wagner, B.; Santini, S.; Ingensand, H.; Gärtner, H. A tool to model 3D coarse-root development with annual resolution. *Plant Soil* **2011**, *346*, 79–96. [CrossRef]
31. Smith, A.; Astrup, R.; Raumonen, P.; Liski, J.; Krooks, A.; Kaasalainen, S.; Åkerblom, M.; Kaasalainen, M. Tree Root System Characterization and Volume Estimation by Terrestrial Laser Scanning and Quantitative Structure Modeling. *Forests* **2014**, *5*, 3274–3294. [CrossRef]
32. Sakamoto, T.; Noguchi, H.; Gotoh, Y.; Suzuki, S.; Shimada, K. Management of Japanese Black Pine (*Pinus thunbergii*) Seedlings Caused by Natural Regeneration. *J. Jpn. Soc. Coast. For.* **2013**, *12*, 29–34, (In Japanese with English summary)
33. Tanaka, J.; Nakazawa, H.; Satou, T. Case Study of Root Systems Growth of *Pinus thunbergii* Parlatore and Groundwater Level in the Nishinohama Coastal Forest, Tahara City, Aichi Prefecture. *J. JSRT* **2017**, *43*, 298–301. (In Japanese) [CrossRef]
34. Hirano, Y.; Todo, C.; Yamase, K.; Tanikawa, T.; Dannoura, M.; Ohashi, M.; Doi, R.; Wada, R.; Ikeno, H. Quantification of the Contrasting Root Systems of *Pinus thunbergii* in Soils with Different Groundwater Levels in a Coastal Forest in Japan. *Plant Soil* **2018**, *426*, 327–337. [CrossRef]
35. Todo, C.; Tokoro, C.; Yamase, K.; Tanikawa, T.; Ohashi, M.; Ikeno, H.; Dannoura, M.; Miyatani, K.; Doi, R.; Hirano, Y. Stability of *Pinus thunbergii* Between Two Contrasting Stands at Differing Distances from the Coastline. *For. Ecol. Manag.* **2019**, *431*, 44–53. [CrossRef]
36. FAO-UNESCO. Soil Map of the World. Revised Legend. Reprinted with Corrections. In *World Soil Resources Report 60*; Food and Agriculture Organization: Rome, Italy, 1990; p. 119.
37. Japan Meteorological Agency. Available online: http://www.data.jma.go.jp/obd/stats/etrn/ (accessed on 18 April 2021).
38. FARO Technologies Inc. FARO Laser Scanner Focus. Available online: https://faro.app.box.com/s/kfpwjofogeegocr7mf2s866s2qalnaqw (accessed on 18 April 2021).
39. Feng, L.; Zhao, T.; Kim, J. neuTube 1.0: A New Design for Efficient Neuron Reconstruction Software Based on the Swc Format. *eNeuro* **2015**, *2*. [CrossRef] [PubMed]
40. Yang, M.; Defossez, P.; Dupont, S. A Root-to-Foliage Tree Dynamic Model for Gusty Winds During Wind Storm Conditions. *Agric. Meteorol.* **2020**, *287*, 107949. [CrossRef]
41. R Core Team. *R: A Language and Environment for Statistical Computing*; R Foundation for Statistical Computing: Vienna, Austria, 2017; Available online: https://www.R-project.org/ (accessed on 12 July 2020).
42. Liski, J.; Kaasalainen, S.; Raumonen, P.; Akujärvi, A.; Krooks, A.; Repo, A.; Kaasalainen, M. Indirect Emissions of Forest Bioenergy: Detailed Modeling of Stump-Root Systems. *GCB Bioenergy* **2014**, *6*, 777–784. [CrossRef]
43. Coutts, M.P. Components of Tree Stability in Sitka Spruce on Peaty Gley Soil. *Forestry* **1986**, *59*, 173–197. [CrossRef]
44. Nicoll, B.C.; Ray, D. Adaptive Growth of Tree Root Systems in Response to Wind Action and Site Conditions. *Tree Physiol.* **1996**, *16*, 891–898. [CrossRef] [PubMed]
45. Nicoll, B.C.; Easton, E.P.; Milner, A.D.; Walker, C.; Coutts, M.P. Wind Stability Factors in Tree Selection: Distribution of Biomass within Root Systems of Sitka Spruce Clones. In *Wind and Trees*; Coutts, M.P., Grace, J., Eds.; Cambridge University Press: Cambridge, UK, 1995; pp. 276–292.

46. Boudon, F.; Preuksakarn, C.; Ferraro, P.; Diener, J.; Nacry, P.; Nikinmaa, E.; Godin, C. Quantitative Assessment of Automatic Reconstructions of Branching Systems Obtained from Laser Scanning. *Ann. Bot.* **2014**, *114*, 853–862. [CrossRef]
47. Hirano, Y.; Yamamoto, R.; Dannoura, M.; Aono, K.; Igarashi, T.; Ishii, M.; Yamase, K.; Makita, N.; Kanazawa, Y. Detection Frequency of *Pinus thunbergii* Roots by Ground-Penetrating Radar Is Related to Root Biomass. *Plant Soil* **2012**, *360*, 363–373. [CrossRef]
48. Kaestner, A.; Schneebeli, M.; Graf, F. Visualizing Three-Dimensional Root Networks Using Computed Tomography. *Geoderma* **2006**, *136*, 459–469. [CrossRef]

forests

MDPI

Article

Root Foraging Ability for Phosphorus in Different Genotypes *Taxodium* 'Zhongshanshan' and Their Parents under Phosphorus Deficiency

Rongxiu Xie [1], Jianfeng Hua [2], Yunlong Yin [2] and Fuxu Wan [1,*]

1 College of Forestry, Nanjing Forestry University, 159 Longpan Road, Xuanwu District, Nanjing 210037, China; RongxiuXie@hotmail.com

2 Jiangsu Engineering Research Center for Taxodium Rich, Germplasm Innovation and Propagation, Institute of Botany, Jiangsu Province and Chinese Academy of Sciences, Nanjing 210014, China; jfhua@cnbg.net (J.H.); yinyl066@sina.com (Y.Y.)

* Correspondence: fxwan@njfu.com.cn

Abstract: The phosphorus (P) deficiency is the one of the key constraints for Taxodium 'Zhongshanshan' afforestation. A hydroponic experiment was conducted to explore root foraging ability for P in different genotypes of *Taxodium* 'Zhongshanshan' (*T.*'Zhongshanshan') and their parents (*T.mucronatum* and *T.distichum*). Five P levels of CK (31 mg/L), P15 (15 mg/L), P10 (10 mg/L), P5 (5 mg/L), and P0 (0 mg/L) were set up as the P deficiency stress treatment. The plant P contents, root morphological indices, and plant growth traits of different taxodium genotypes were measured. Meanwhile, the root foraging ability for P was evaluated with the membership function method in combination with weight. Results showed that: (1) Except the plant P content, the root morphology, plant net biomass, and height showed significant differences among the different genotypes ($p < 0.05$); the P deficiency stress had no significant influence on root morphology, but a significant influence on plant net biomass and height and P content; (2) *T.mucronatum* and *T.*'Zhongshanshan'302 had relatively lower values of root length, root surface area, root volume, and plant net biomass, but had no difference of plant P content with the other genotypes; (3) *T.mucronatum* and *T.*'Zhongshanshan'302 had higher root foraging ability for P than the other genotypes; (4) the stepwise regression analysis revealed the root volume as the main factor significantly influencing the root foraging ability. This study concluded that different genotypes of *T.*'Zhongshanshan' and their parents had different root foraging ability for P, and breeding and screening the fine varieties is conducive for the afforestation in P-limited areas.

Keywords: phosphorus deficiency; *T.*'Zhongshanshan'; root foraging ability for phosphorus

Citation: Xie, R.; Hua, J.; Yin, Y.; Wan, F. Root Foraging Ability for Phosphorus in Different Genotypes *Taxodium* 'Zhongshanshan' and Their Parents under Phosphorus Deficiency. *Forests* **2021**, *12*, 215. https://doi.org/10.3390/f12020215

Academic Editor: Antonio Montagnoli

Received: 27 December 2020
Accepted: 6 February 2021
Published: 12 February 2021

1. Introduction

Phosphorus (P) is an essential element for plant growth and crop production [1,2]. It plays a significant role in metabolic processes related to cellular processes involved in generating and transforming metabolic energy [3,4]. However, large areas of tropical and subtropical soils in Africa, Latin America, and Asia have P availability limited by low total P content as well as high P fixation in soils [5,6], thus inhibiting agricultural and forestry productivity [7]. In the past decades, P fertilizer was commonly used to improve soil P deficiency and obtain high yields. However, P fertilizers are costly and potentially harmful to the environment, such as water eutrophication and soil P enrichment [8,9]. Thus, a genetic screen of plant low P tolerance is an important strategy to increase agricultural and forestry productivity to satisfy the demand of P in plants and reduce the amount of P fertilizers [10].

Roots play a crucial role as the primary path for P uptake by plants. P uptake depends mainly on root characteristics [11]. Da Silva A (2015) reported that twenty-one Brazilian wheat cultivars showed potentially greater phosphorus uptake efficiency, and we observed the importance of root traits for improving the P uptake ability [12]. Previous studies have shown that plants with low P tolerance were associated with root foraging ability for P, and the ability was different due to the genotype differences, even for the same species [13–17]. For example, Wu (2019) reported that P deficiency resulted in root proliferation of Chinese Fir (*Cunninghamia lanceolata*), including increases in root length, root volume, biomass of root, and root-shoot ratio, which contributes to foraging P. Besides, families No. 25, 20, and 41 were elite Chinese Fir with strong P foraging ability through comprehensive evaluation on root morphology and architecture [18]. Aziz T (2011) reported that Two Brassica cultivars for growth and P uptake was associated with their longer roots [19]. Thus, exploring the differences of root foraging ability for P in among different genotypes under P deficiency had an important effect on selecting elite germplasm resources with strong P foraging ability.

Taxodium 'Zhongshanshan' (*T.*'Zhongshanshan') is an interspecies hybrid of *T.distichum* and *T.mucronatum*. It has been widely planted in the coastal and wetland areas of south-eastern China due to its great ecological and economic potential [20–23]. Currently, there have been increasing concerns on *T.*'Zhongshanshan' afforestation adaptability, growth rules, salt resistance, cold resistance, and flooding tolerance [24–27]. For example, *T.*'Zhongshanshan'302 (*T.distichum*♀× *T.mucronatum*♂) is well-adapted to grow in saline environment, while *T.*'Zhongshanshan'118 (*T.*'Zhongshanshan'302♀× *T.mucronatum*♂) is well-adapted to grow the waterlogging, and *T.*'Zhongshanshan'406 (*T.mucronatum*♀× *T.distichum*♂) have been widely used as timber trees [28–30]. However, the effects of P deficiency on root foraging ability for P in different genotypes of *T.*'Zhongshanshan' and their parents have not been well studied.

To fill this gap, we established a hydroponics experiment in the Institute of Botany, Jiangsu Province, and Chinese Academy of Sciences, Nanjing, Jiangsu province, China. In this study, we examine the plant P contents (P content of whole plant, P content of above-ground, and P content of underground), root morphology (root length, root surface area, root volume) *T.*'Zhongshanshan'and plant growth traits (biomass, basal diameter, plant height, root-shoot ratio) of different genotypes *T.* 'Zhongshanshan' (*T.*'Zhongshanshan'118, *T.*'Zhongshanshan'302,*T.*'Zhongshanshan'406) and their parents (*T.mucronatum* and *T.distichum*) under P deficiency. The main objectives of this study were (1): To investigate the effects of P deficiency on plant P contents, root morphology, and plant growth traits of different genotypes and their parents, (2) to evaluate root foraging ability for P in different genotypes *T.*'Zhongshanshan' and their parents. This study provides important knowledge for the further P fertilization management in *T.*'Zhongshanshan' and the selection of elite *T.*'Zhongshanshan' resources with strong P foraging ability.

2. Materials and Methods

2.1. Experimental Materials

The hydroponics experiment was conducted in a greenhouse of the Institute of Botany, Jiangsu Province, and the Chinese Academy of Sciences (118°49′ E, 32°03′ N), with a temperature of 18 °C–28 °C, relative humidity >80%, and day photoperiod of 10h. One-year-old healthy cuttings (*T.*'Zhongshanshan'118,*T.*'Zhongshanshan'302, *T.*'Zhongshanshan'406, *T.mucronatum* and *T.distichum*) were selected. Their basal diameter, plant height, and biomass were determined as follows (Table 1). Table 1 Basal diameter, plant height, and biomass for *Taxodium* plants.

2.2. Experimental Design

Prior to the hydroponics experiment, one-year-old healthy cuttings were carefully transplanted into plastic pots (40cm × 30cm × 25cm), three plants per pot. The dosage of five P levels was designed as 31mg/L (Normal P supply, CK), 15 mg/L (Mild P stress, P15), 10 mg/L (Moderate P stress, P10), 5 mg/L (Severe P stress, P5), 0 mg/L (Extreme P stress, P0), with each group consisting of 5 replicates, totaling to 125 pots.

Table 1. Basal diameter, plant height, and biomass for *Taxodium* plants.

Plants	Basal Diameter/mm	Plant Height/cm	Biomass/g
T.mucronatum	3.21 ± 0.81	18.42 ± 2.7	1.87 ± 0.6
T.distichum	4.26 ± 0.9	23.31 ± 4.25	3.6 ± 1.76
T.'Zhongshanshan'118	3.39 ± 0.69	13.33 ± 3.75	2.42 ± 1.31
T.'Zhongshanshan'302	4.37 ± 1.00	19.07 ± 3.19	3.24 ± 1.79
T.'Zhongshanshan'406	4.68 ± 1.19	23.18 ± 4.22	4.82 ± 1.72
Mean value	4 ± 1.09	19.54 ± 5.13	3.21 ± 1.81

The P solution of the designated concentration was made by mixing the KH_2PO_4 with modified Hoagland's nutrient solution [31]. To compensate for the difference in K supplied, KCl was added to the low-P treatment. The basal modified Hoagland's nutrient solution had the following composition: KNO_3 (5 mM), $Ca(NO_3)_2$ (5 mM), $MgSO_4$ (2 mM), Fe-EDTA (1 mM), H_3BO_3 (46.3 μM), $MnCl_2.4H_2O$ (9.1 μM), $ZnSO_4·7H_2O$ (0.8 μM), $CuSO_4·5H_2O$ (0.3 μM), $H_2MoO_4·4H_2O$ (0.38 μM). The nutrient solution was renewed every 14d and adjusted to pH 6.0 with NaOH (1 M) or HCl (1 M).

2.3. Plants Measurements

The plant height, basal diameter, and biomass were measured before and after the hydroponics experiment. After hydroponics experiment, roots of *Taxodium* plants were harvested separately. The root length, root volume, and root surface area were quantified by the root scanning system (WinRHIZO, version 4.0b). After that, roots and shoots were dried at 80 °C until constant weight to derive their dry weights and the ratio of root–shoot ratio. The plant P contents was determined spectrophotometrically using ammonium molybdate blue method.

2.4. Statistical Analysis

Results were expressed as means ± standard errors. All statistical procedures were conducted using SPSS 21.0 (SPSS software, version 21.0, Inc., Chicago, IL, USA). Two-way ANOVA was used to test the main effects and interactions of genotypes and P levels on plant P contents, root morphology, and plant growth. Correlation analysis was performed to reveal the relationships among the above traits. The membership function method in combination with weight was performed to evaluate root foraging ability for P in different genotypes under P deficiency. The main factors driving changes of root foraging ability for phosphorus were identified by using the best statistical model suggested by stepwise regression in this study. In order to more reasonably evaluate the sensitivity of *Taxodium* plants to low P stress, correlation analysis, membership function method, and regression analysis all used the ratio between each indicator of the low P treatment and the control group as its index value.

3. Results

3.1. Interactive Effects of Experimental Materials and P Levels

Except the P content of the whole plant and aboveground, the P content of underground, root morphology, and growth showed significant differences among experimental genotypes ($p < 0.05$). Root surface area, biomass, plant height, root–shoot ratio (fresh weight), root–shoot ratio (dry weight), P content of aboveground, P content of underground, and P content of whole plant showed significant differences ($p < 0.05$) among different P levels, while no significant differences were found in root length, root volume, and net basal diameter. Moreover, there were significant interaction effects of genotypes and P levels on the net basal diameter, net plant height, root–shoot ratio (fresh weight and dry weight), P content of aboveground, P content of underground among experimental materials, and P levels ($p < 0.05$) (Table 2).

Table 2. *F*-value and the results of two-way ANOVA testing the different genotypes and P levels on the growth indexes of *Taxodium* plants.

Indicators	Materials		P Level		Materials × P Level	
	F	*p*	*F*	*p*	*F*	*p*
Root length	32.720	0.000	2.248	0.070	0.637	0.846
Root surface area	26.770	0.000	2.668	0.037	0.703	0.784
Root volume	27.036	0.000	0.886	0.476	1.232	0.259
Net biomass	13.512	0.000	7.496	0.000	1.308	0.209
Net basal diameter	7.316	0.000	1.284	0.282	2.106	0.014
Net plant height	5.103	0.001	15.109	0.000	2.571	0.002
Root-shoot ratio (fresh weight)	12.724	0.000	32.825	0.000	3.719	0.000
Root-shoot ratio (dry weight)	14.362	0.000	25.047	0.000	5.179	0.000
P content of aboveground	2.451	0.051	14.717	0.000	1.315	0.203
P content of underground	3.076	0.020	12.224	0.000	2.249	0.008
P content of whole plant	1.528	0.200	5.373	0.001	2.010	0.019

P level = Phosphorus level, F = F Value, *p* = Significance.

3.2. Changes of Root Morphology

No significant effects on root morphology of *T.distichum*, *T.*'Zhongshanshan'118, and *T.*'Zhongshanshan'302 were found in this study (Figure 1). However, the mean values of root length and root surface area of *T.*'Zhongshanshan'406 were significantly higher ($p < 0.05$) in the P15 treatment than those in the P5 treatment (Figure 1a,b). The mean value of root volume of *T.mucronatum* was significantly higher ($p < 0.05$) in the P5 treatment than that in the P0 treatment (Figure 1c).

There were significant effects of P levels on the root length, root surface area, and root volume of *T.mucronatum* when its root diameter was in the range of 0.5~1.0 mm. Besides, significant effects of P levels on the root length of *T.*'Zhongshanshan'406 was observed when its root diameter was from 0~0.5 mm (Table 3).

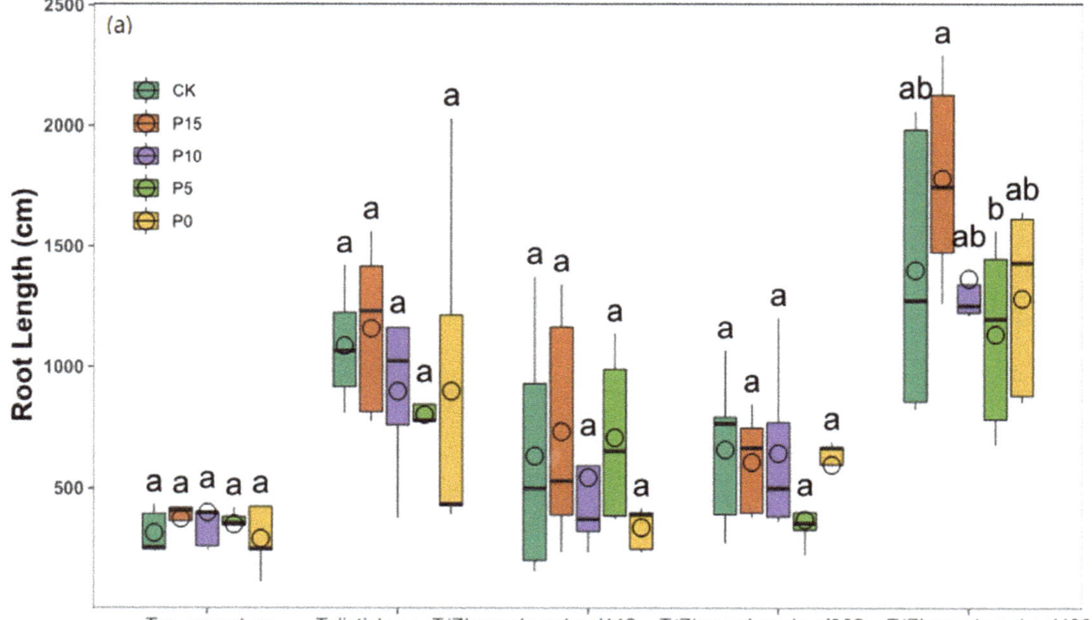

Figure 1. *Cont.*

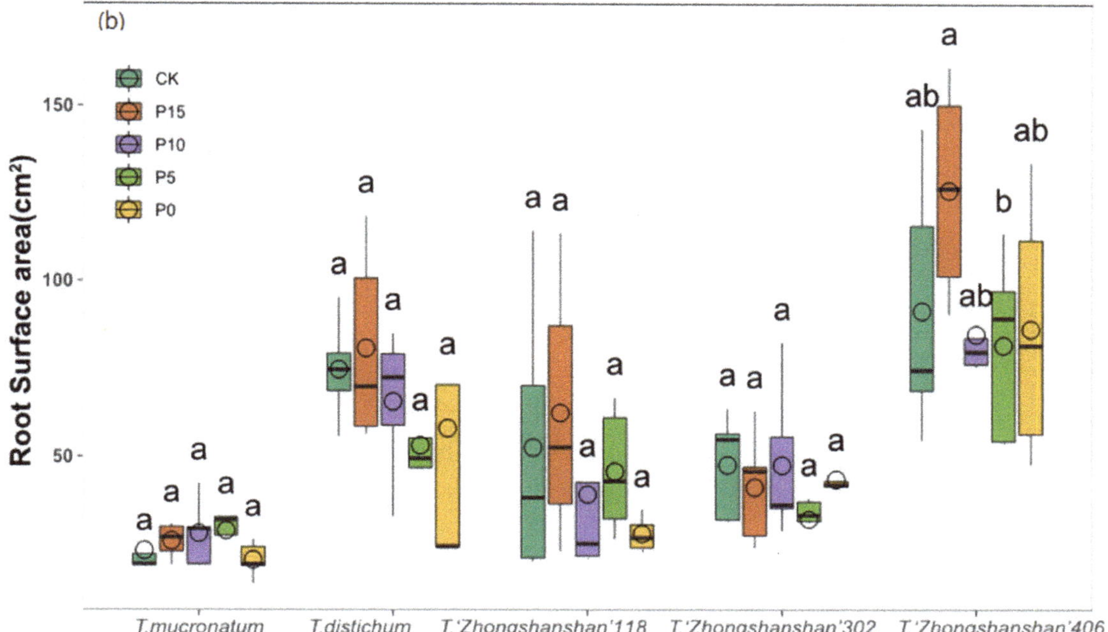

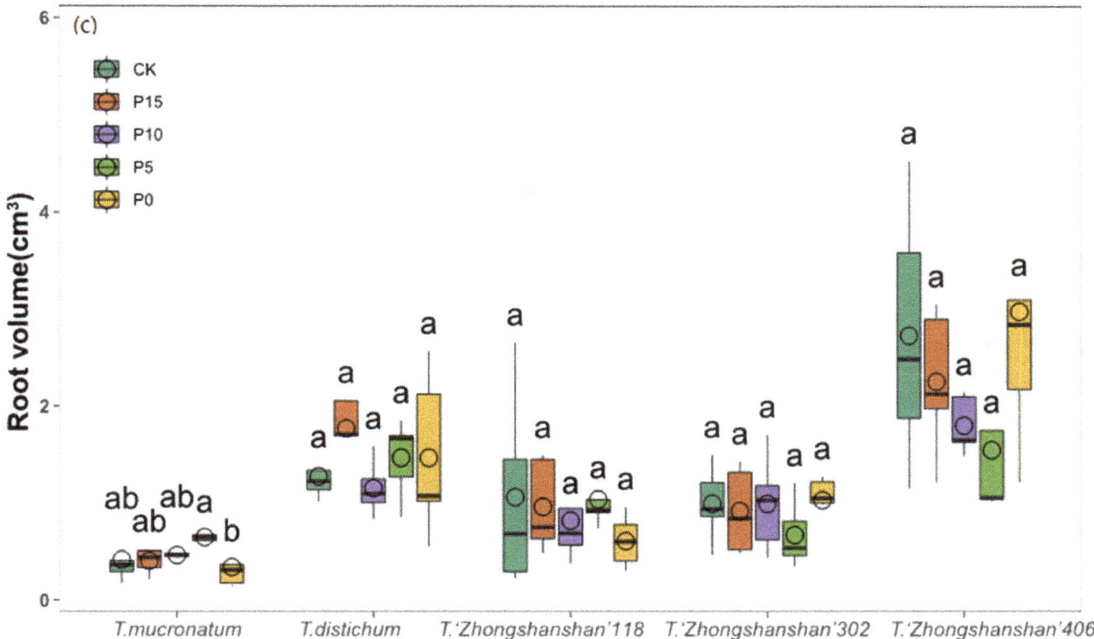

Figure 1. Root length (**a**), root surface area (**b**), and root volume (**c**) of *Taxodium* plants under P deficiency experiment. CK = Normal P supply, P15 = Mild P stress, P10 = Moderate P stress, P5 = Severe P stress, P0 = Extreme P stress, P0.

Table 3. Effect of phosphorus level on different root length ranges, different root surface area ranges, and different root volume ranges of *Taxodium* plants.

Indicators	Measuring Range	*T.Mucronatum*		*T.Distichum*		*T.*'Zhongshanshan'118		*T.*'Zhongshanshan'302		*T.*'Zhongshanshan'406	
		F	*p*	*F*	*p*	*F*	*p*	*F*	*p*	*F*	*p*
Root length (cm)	0 < Root length ≤ 0.500	0.593	0.672	0.973	0.445	1.29	0.311	2.086	0.121	3.609	0.023
	0.500 < Root length ≤ 1.000	3.900	0.018	0.718	0.590	0.320	0.861	0.279	0.888	1.467	0.250
	1.000 < Root length ≤ 1.500	1.548	0.229	0.988	0.438	0.696	0.604	1.589	0.216	1.070	0.398
	1.500 < Root length ≤ 2.000	0.72	0.589	2.13	0.117	0.624	0.651	1.449	0.255	2.171	0.109
	2.000 < Root length ≤ 2.500	0.787	0.548	1.639	0.206	0.727	0.585	0.878	0.495	1.808	0.167
	2.500 < Root length ≤ 3.000	0.173	0.949	2.068	0.125	0.607	0.662	1.022	0.420	1.173	0.352
	Root length > 3.000	0.668	0.622	1.632	0.207	0.785	0.550	0.986	0.438	1.807	0.167
Root surface area (cm^2)	0 < Root surface area ≤ 0.500	0.702	0.600	0.929	0.468	1.422	0.267	1.952	0.141	2.82	0.053
	0.500 < Root surface area ≤ 1.000	3.829	0.019	0.743	0.574	0.317	0.863	0.275	0.890	1.418	0.264
	1.000 < Root surface area ≤ 1.500	1.435	0.261	0.963	0.45	0.698	0.604	1.611	0.210	1.168	0.355
	1.500 < Root surface area ≤ 2.000	0.731	0.582	2.125	0.117	0.644	0.638	1.449	0.255	2.140	0.113
	2.000 < Root surface area ≤ 2.500	0.778	0.553	1.646	0.204	0.716	0.592	0.851	0.510	1.798	0.169
	2.500 < Root surface area ≤ 3.000	0.161	0.955	2.019	0.133	0.598	0.669	1.002	0.430	1.179	0.350
	Root surface area > 3.000	0.636	0.643	1.648	0.204	0.757	0.567	0.989	0.436	1.785	0.171
Root volume (cm^3)	0 < Root volume ≤ 0.500	0.941	0.462	0.892	0.488	1.383	0.279	1.827	0.163	2.105	0.118
	0.500 < Root volume ≤ 1.000	3.717	0.021	0.76	0.564	0.315	0.864	0.291	0.881	1.352	0.286
	1.000 < Root volume ≤ 1.500	1.319	0.299	0.935	0.465	0.698	0.603	1.634	0.205	1.276	0.312
	1.500 < Root volume ≤ 2.000	0.743	0.575	2.117	0.118	0.663	0.625	1.453	0.254	2.110	0.117
	2.000 < Root volume ≤ 2.500	0.767	0.560	1.660	0.201	0.706	0.598	0.836	0.519	1.786	0.171
	2.500 < Root volume ≤ 3.000	0.151	0.960	1.968	0.140	0.588	0.675	0.988	0.437	1.184	0.348
	Root volume > 3.000	0.603	0.665	1.679	0.196	0.736	0.580	0.993	0.434	1.762	0.176

F = F Value, p = Significance.

3.3. Changes of Plant Growth

P deficiency led to significant effects on plant growth of *T.mucronatum*, *T.distichum*, *T.*'Zhongshanshan'118, *T.*'Zhongshanshan'302, and *T.*'Zhongshanshan'406 ($p < 0.05$) (Figure 2). The mean values of net biomass and net plant height of *T.mucronatum* were significantly lower in the P0 treatment than those in the CK treatment (Figure 2a,c), while the mean value of net basal diameter of *T.mucronatum* was significantly higher in the P15 treatment than that in the CK treatment (Figure 2b). The mean value of net plant height of *T.distichum* was significantly lower ($p < 0.05$) in the P0 treatment than that in the CK treatment (Figure 2c). The mean values of net biomass and net plant height of *T.*'Zhongshanshan'118 were significantly lower in the P0 treatment than those in the CK treatment (Figure 2a,c). The mean value of net basal diameter of *T.*'Zhongshanshan'302 was significantly lower ($p < 0.05$) in the P deficiency treatments than that in the CK treatment (Figure 2b). The mean values of net biomass and net plant height of *T.*'Zhongshanshan'406 were significantly lower ($p < 0.05$) in the P deficiency treatments than those in the CK treatment (Figure 2a,c).

3.4. Changes of Root–Shoot Ratio

P deficiency led to significant effects on root–shoot ratio of *T.mucronatum*, *T.distichum*, *T.*'Zhongshanshan'118, *T.*'Zhongshanshan'302, and *T.*'Zhongshanshan'406 ($p < 0.05$) (Figure 3). The mean values of root–shoot ratio (dry weight) and root–shoot ratio (fresh weight) of *T.mucronatum* were significantly higher ($p < 0.05$) in the P10, P5, and P0 treatments than those in the CK treatment (Figure 3a,b). The mean values of root–shoot ratio (dry weight) of *T.distichum* were significantly higher ($p < 0.05$) in the P10 and P0 than those in the CK treatment, the mean value of root–shoot ratio (fresh weight) of *T.distichum* was significantly higher ($p < 0.05$) in the P0 treatment than that in the CK treatment (Figure 3a,b). The mean values of root–shoot ratio (dry weight) and root–shoot ratio (fresh weight) of *T.*'Zhongshanshan' 118 were significantly higher ($p < 0.05$) in the P0 treatment than those in the CK treatment (Figure 3a,b). The mean value of root–shoot ratio (fresh weight) of *T.*'Zhongshanshan'302 was significantly higher ($p < 0.05$) in P0 than that in the CK treatment (Figure 3a,b). The mean values of root–shoot ratio (dry weight) of *T.*'Zhongshanshan'406 were significantly higher ($p < 0.05$) in P15 and P0 than those in the CK treatment, the mean value of root–shoot ratio (fresh weight) of *T.*'Zhongshanshan' 406 was significantly higher ($p < 0.05$) in the P0 treatment than that in CK (Figure 3a,b).

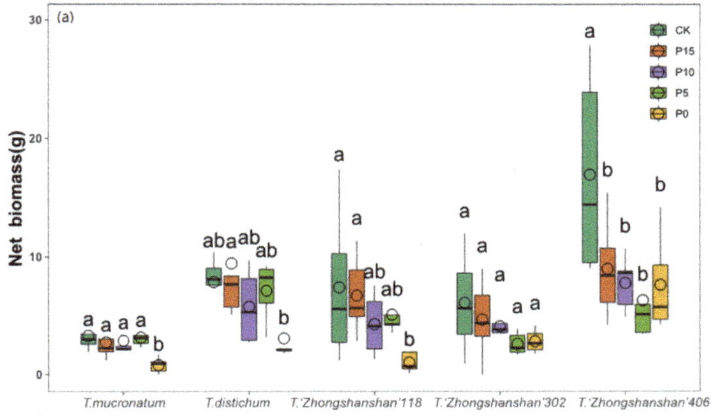

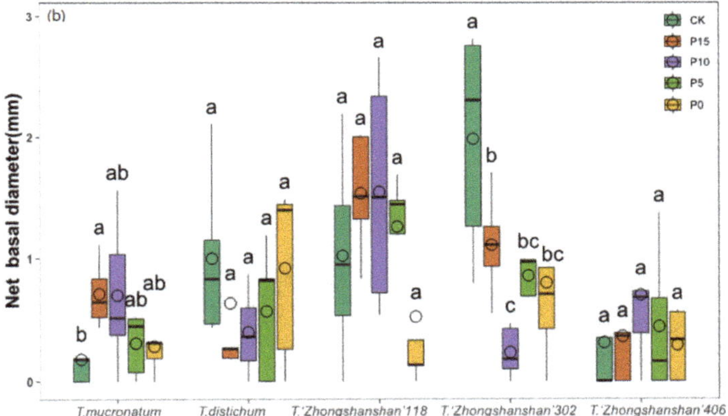

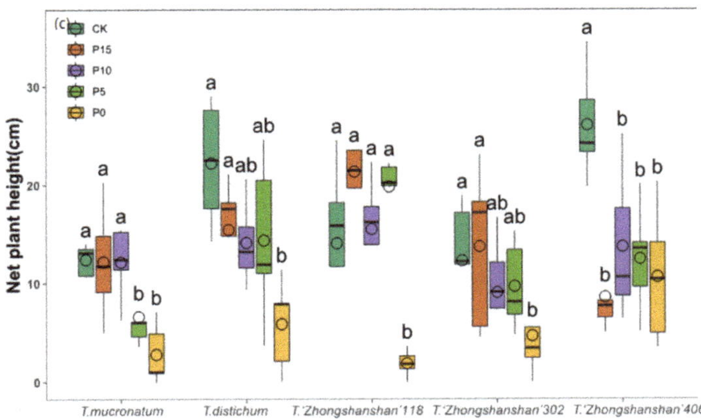

Figure 2. Net biomass (**a**), net basal diameter (**b**) and net plant height (**c**) of *Taxodium* plants under P deficiency experiment. CK = Normal P supply, P15 = Mild P stress, P10 = Moderate P stress, P5 = Severe P stress, P0 = Extreme P stress, P0.

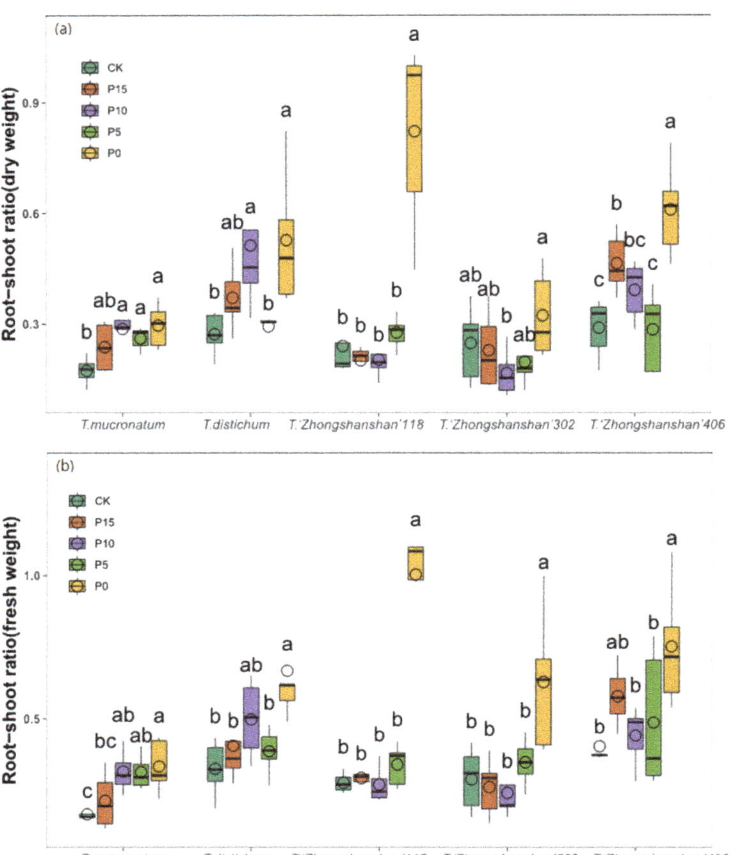

Figure 3. Root–shoot ratio (dry weight) (**a**) and root–shoot ratio (fresh weight) (**b**) of *Taxodium* plants under P deficiency experiment. CK = Normal P supply, P15 = Mild P stress, P10 = Moderate P stress, P5 = Severe P stress, P0 = Extreme P stress, P0.

T.distichum, *T.*'Zhongshanshan'118, *T.*'Zhongshanshan'302, and *T.*'Zhongshanshan' 406 ($p < 0.05$) (Figure 3). The mean values of root–shoot ratio (dry weight) and root–shoot ratio (fresh weight) of *T.mucronatum* were significantly higher ($p < 0.05$) in the P10, P5, and P0 treatments than those in the CK treatment (Figure 3a,b). The mean values of root–shoot ratio (dry weight) of *T.distichum* were significantly higher ($p < 0.05$) in the P10 and P0 than those in the CK treatment, the mean value of root–shoot ratio (fresh weight) of *T.distichum* was significantly higher ($p < 0.05$) in the P0 treatment than that in the CK treatment (Figure 3a,b). The mean values of root–shoot ratio (dry weight) and root–shoot ratio (fresh weight) of *T.*'Zhongshanshan' 118 were significantly higher ($p < 0.05$) in the P0 treatment than those in the CK treatment (Figure 3a,b). The mean value of root–shoot ratio (fresh weight) of *T.*'Zhongshanshan'302 was significantly higher ($p < 0.05$) in P0 than that in the CK treatment (Figure 3a,b). The mean values of root–shoot ratio (dry weight) of *T.*'Zhongshanshan'406 were significantly higher ($p < 0.05$) in P15 and P0 than those in the CK treatment, the mean value of the root–shoot ratio (fresh weight) of *T.*'Zhongshanshan' 406 was significantly higher ($p < 0.05$) in the P0 treatment than that in CK (Figure 3a,b).

3.5. Changes of Plant P Contents

P content of aboveground for *T.mucronatum* was significantly lower ($p < 0.05$) in the P0 treatment than that in the P5 treatment (Figure 4b), P content of underground for *T. mucronatum* were significantly lower ($p < 0.05$) in the P treatments than that in the CK treatment (Figure 4c). P content of whole plant and P content of aboveground for *T.distichum* were significantly lower ($p < 0.05$) in the P10, P5, and P0 treatments than those in the CK treatment (Figure 4a,b). P content of whole plant and P content of aboveground for *T.*'Zhongshanshan'118 were significantly lower ($p < 0.05$) in the P0 treatment than those in the CK treatment (Figure 4a,b). P content of whole plant for *T.*'Zhongshanshan'302 were significantly lower ($p < 0.05$) in the P treatments than those in the CK treatment (Figure 4a), P content of aboveground for *T.*'Zhongshanshan'302 were significantly lower ($p < 0.05$) in P10 and P5 treatments than those in the CK treatment (Figure 4b), P content of underground for *T.*'Zhongshanshan'302 were significantly lower ($p < 0.05$) in P0 treatment than that in the CK treatment (Figure 4c). P content of whole plant and P content of aboveground for *T.*'Zhongshanshan'406 were significantly lower ($p < 0.05$) in the P treatments than those in the CK treatment (Figure 4a,b).

3.6. Correlation Analysis

Correlation analysis (using SPSS 21.0) showed that net plant height had significant positive correlation with net biomass ($p < 0.01$), P content of whole plant ($p < 0.05$), and P content of aboveground ($p < 0.05$), while net plant height had significant negative correlation with root–shoot ratio (fresh weight), root–shoot ratio (dry weight) ($p < 0.01$). Net biomass had significant positive correlation with root volume, root surface area, P content of whole plant, and P content of aboveground ($p < 0.01$), while net biomass had significant negative correlation with root–shoot ratio (fresh weight), root–shoot ratio (dry weight) ($p < 0.01$). Net basal diameter had significant positive correlation with root surface area ($p < 0.05$), while net basal diameter had significant negative correlation with P content of underground ($p < 0.01$). Root volume had significant positive correlation with root surface area ($p < 0.05$) and P content of aboveground ($p < 0.01$). Root surface area and root length had significant positive correlation with P content of aboveground ($p < 0.05$). Root length had significant positive correlation with P content of whole plant ($p < 0.01$). Root–shoot ratio (fresh weight) had significant positive correlation with root–shoot ratio (dry weight) ($p < 0.01$), while root–shoot ratio (fresh weight) significant negative correlation with P content of whole plant ($p < 0.01$), P content of aboveground ($p < 0.05$). Root–shoot ratio (dry weight) significant negative correlation with P content of aboveground ($p < 0.05$), P content of whole plant ($p < 0.01$). P content of whole plant had significant positive correlation with P content of underground ($p < 0.01$), P content of underground ($p < 0.05$) (Table 4).

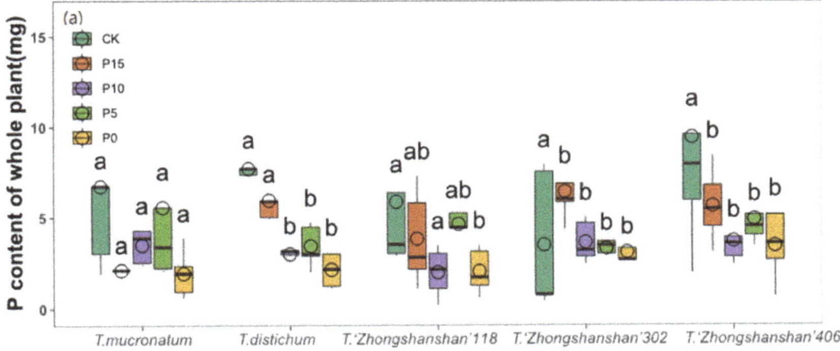

Figure 4. *Cont.*

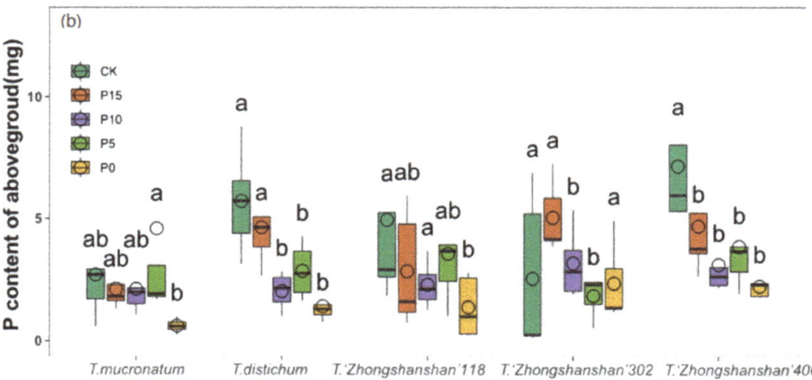

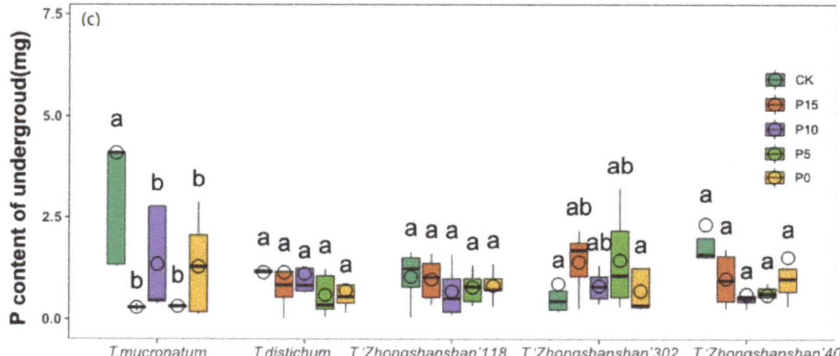

Figure 4. P content of whole plant (**a**), P content of aboveground (**b**), and P content of underground (**c**) for *Taxodium* plants under P deficiency experiment. CK = Normal P supply, P15 = Mild P stress, P10 = Moderate P stress, P5 = Severe P stress, P0 = Extreme P stress, P0.

Table 4. Correlation matrix of every single index.

Indicators	Net Plant Height	Net Biomass	Net Basal Diameter	Root Volume	Root Surface Area	Root Length	Root–Shoot Ratio (Fresh Weight)	Root–Shoot Ratio (Dry Weight)	P Content of Whole Plant	P Content of Above-ground	P Content of Underground
Net plant height	1										
Net biomass	0.635 **	1									
Net basal diameter	0.206	0.114	1								
Root volume	0.097	0.634 **	0.050	1							
Root surface area	0.228	0.515 **	0.449 *	0.453 *	1						
Root length	−0.208	0.200	−0.390	0.374	0.077	1					
Root–shoot ratio (fresh weight)	−0.648 **	−0.560 **	0.001	−0.044	−0.289	−0.248	1				
Root–shoot ratio (dry weight)	−0.651 **	−0.509 **	0.041	−0.099	−0.245	−0.313	0.913 **	1			
P content of whole plant	−0.492 *	0.746 **	−0.120	0.371	0.298	0.292	−0.530 **	−0.561 **	1		
P content of aboveground	0.398 *	0.753 **	0.216	0.595 **	0.515 **	0.469 *	−0..345	−0.416 *	0.828 **	1	
P content of underground	0.386	0.391	−0.506 **	0.088	−0.176	−0.112	−0.402 *	−0.274	0.501 *	0.039	1

Note: ** Correlation is significant at $\alpha = 0.01$ (2 − tailed). * Correlation is significant at $\alpha = 0.05$ (2−tailed).

3.7. Comprehensive Evaluation of Root P-Foraging Ability

The membership function method in combination with weight [22,32] was performed to evaluate root foraging ability for P in different genotypes *T.*'Zhongshanshan' and their parents under P deficiency, and gaining the final result shown as Table 5. Results showed that the comprehensive evaluation value (D) ranged from 0.599 for *T.mucronatum* to 0.351 for *T.*'Zhongshanshan'406, besides, comprehensive evaluation value for *T.*'Zhongshanshan'302 was the highest among the three *T.* 'Zhongshanshan' genotypes. These result indicated that *T. mucronatum* and *T.*'Zhongshanshan'302 had higher root foraging ability for P than *T.distichum*, *T.*'Zhongshanshan'118 and *T.*'Zhongshanshan'406 (Table 5).

Table 5. Comprehensive evaluation on growth indexes of *Taxodium* plants under P deficiency.

Plants	Evaluation *D*				
	P15	P10	P5	P0	Mean Value (D)
T.mucronatum	0.516	0.659	0.708	0.513	0.599
T.distichum	0.588	0.305	0.326	0.571	0.448
T.'Zhongshanshan'118	0.481	0.560	0.605	0.108	0.439
T.'Zhongshanshan'302	0.424	0.443	0.268	0.738	0.468
T.'Zhongshanshan'406	0.210	0.303	0.265	0.627	0.351

P deficiency = Phosphorus deficiency, P15 = Mild P stress, P10 = Moderate P stress, P5 = Severe P stress, P0 = Extreme P stress.

3.8. Stepwise Regression Analysis

Referring to the relationship between totally 11 variables (plant P contents, root morphological, and plant growth traits) and root foraging ability for P, this study set up the stepwise regression model (using SPSS 21.0) of influencing factors of comprehensive evaluation value (*D*) for different genotypes *T.*'Zhongshanshan' and their parents. The best statistical model suggested by stepwise regression in this study was $D = 0.15 + 0.501$ $X1$ ($R2 = 0.502$, F = 18.157, $p = 0.00$), where $X1$ was root volume (Figure 5). This results indicated that root volume was the main factor driving changes of root foraging ability for P in different genotypes *T.* 'Zhongshanshan' and their parents. Besides, the statistical model $D = 0.15 + 0.501\ X1$ could be used for further prediction of root foraging ability for P in different genotypes *T.*'Zhongshanshan' and their parents.

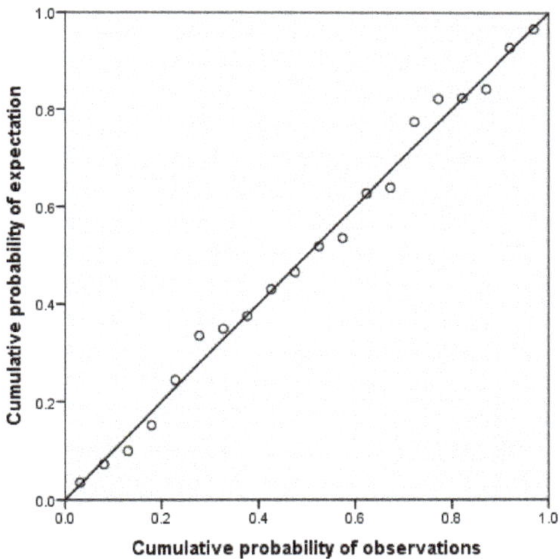

Figure 5. Regression standard P-P graph of normalized residuals (Dependent variable: D).

4. Discussions

Two-way ANOVA showed that P deficiency led to significant effects on root surface area, net biomass, net plant height, root–shoot ratio (fresh weight), root–shoot ratio (dry weight), P content of aboveground, P content of underground, and P content of whole plant (Table 2). Especially in the P0 treatment, P deficiency led to no significant effects on root length, root surface area, and root volume of *T.mucronatum*, *T.distichum*, *T.*'Zhongshanshan'118, *T.*'Zhongshanshan'302, and *T.*'Zhongshanshan'406 ($p > 0.05$, vs. the control treatment) (Figure 1), while the root–shoot ratio (dry weight) and root–shoot ratio (fresh weight) in the P0 treatment were significantly higher than those in the CK treatment (Figure 3). These results suggested that the experimental materials could increase the portion of root and the interaction between the root system and soil, which could improve the P foraging capacity and maintain the growth of the aboveground biomass [33,34].

In the comprehensive evaluation analysis, comprehensive evaluation value for *T.*'Zhongshanshan'302 was the highest among the three *T.*'Zhongshanshan' genotypes, then *T.* 'Zhongshanshan'118 (Table 5). This study also found that the root length, root surface area, and root volume for *T.* 'Zhongshanshan'302 in the P0 treatment were higher than those in the P0 treatment for *T.*'Zhongshanshan'118 (Figure 1). These results indicated that *T.*'Zhongshanshan'302 tend to transfer more photosynthetic product to root system, which increased the growth of root, improving the P foraging capacity. Besides, *T.*'Zhongshanshan'302 is an interspecies hybrid clone generated from *T.distichum*♀, *T.mucronatum*♂, *T.mucronatum* had higher root foraging ability for P than other experimental materials, this result suggested that the low P tolerance of *T.* 'Zhongshanshan'302 was attributed to the inheritance of P deficiency resistance in *T.mucronatum*, which needs further exploration.

T.'Zhongshanshan'406 (*T.mucronatum*♀× *T.distichum*♂) have been selected and demonstrated improvements in growth rate, salt tolerance, form, and vigor [35,36]. However, *T.* 'Zhongshanshan'406 had higher root foraging ability for P than *T.*'Zhongshanshan'302 and *T.*'Zhongshanshan'118. This result may be due to the following two reasons.

Firstly, P deficiency may lead to inhibition effects on photosynthetic pigments synthesis of *T.*'Zhongshanshan'406. Some studies showed that under low phosphorus stress, photosynthetic pigments were inhibited [37], and there was no difference between genotypes. The decrease of photosynthetic parameters of P efficient genotype was smaller than that of p inefficient genotype [38]. The harvested leaves color of *T.*'Zhongshanshan'406 were lighter than those leaves color of *T.*'Zhongshanshan'118 and *T.* 'Zhongshanshan'302, which suggested that the photosynthetic pigments synthesis of *T.*'Zhongshanshan'406 was inhibited under P deficiency. The decreases of photosynthetic pigments synthesis may result in decreasing of photosynthesis, inhibiting the accumulation of dry matter. In this study, the net biomass of *T.*'Zhongshanshan'406 in P treatments was significantly lower than that in the CK treatment (Figure 2a). The decreases of net biomass may lead to decreases in root length and root surface area (Figure 1), which in turn inhibited the P foraging capacity of *T.* 'Zhongshanshan'406.

Secondly, P deficiency may lead to inhibition effects on the root exudates of organic acids in *T.*'Zhongshanshan'406. pH change caused by organic acid secretion from roots is one of the main factors that determine the phosphorus forms in soil [39–41]. Root exudation of organic acids under P deficiency led to decreasing the rhizosphere pH, thus making P more available for plant uptake. In this study, pH test during the hydroponics experiment showed that the mean value of pH in *T.*'Zhongshanshan'406 was higher than that in *T.*'Zhongshanshan'118 and *T.*'Zhongshanshan'302. These results indicated that the low concentration of organic acids of *T.* 'Zhongshanshan'406 under P deficiency may lead to inhibition effects on P availability for plant uptake, which in turn inhibited P foraging capacity of *T.* 'Zhongshanshan'406.

5. Conclusions

In conclusion, *Taxodium* species adapted to P deficiency mainly through the changes of root morphological, such as the significant increases of root–shoot ratio. Besides, *T.mucronatum* and *T.*'Zhongshanshan'302 with higher root foraging ability are suitable for planting in P-limited areas.

Stepwise regression analysis showed that root volume was the main factor driving changes of root foraging ability for phosphorus in different genotypes *T.*'Zhongshanshan' and their parents, besides, root volume was suitable for the statistical model $D = 0.15 + 0.501\ X1$ ($X1$ is root volume), which could be used for further prediction of root foraging ability for P in different genotypes *T.*'Zhongshanshan' and their parents.

Author Contributions: Formal analysis, R.X.; funding acquisition, Y.Y., F.W.; investigation, R.X.; supervision, J.H., Y.Y., and F.W.; writing—original draft, R.X.; writing—review and editing, R.X., J.H., Y.Y., F.W. All authors have read and agreed to the published version of the manuscript.

Funding: This research was funded by A Project Funded by the Priority Academic Program Development of Jiangsu Higher Education Insitutions (PAPD), Biological Resources Service Network (kfj-brsn-2018-6-003).

Institutional Review Board Statement: Not applicable.

Informed Consent Statement: Not applicable.

Data Availability Statement: Data available on request due to restrictions of privacy. The data presented in this study are available on request from the corresponding author.

Acknowledgments: We thank Chaoguang Yu and Ruirui Li for their help in the research process.

Conflicts of Interest: The authors declare no conflict of interest.

References

1. Kvakić, M.; Pellerin, S.; Ciais, P.; Achat, D.L.; Augusto, L.; Denoroy, P.; Gerber, J.S.; Goll, D.; Mollier, A.; Mueller, N.D.; et al. Quantifying the Limitation to World Cereal Production Due To Soil Phosphorus Status. *Glob. Biogeochem. Cycles* **2018**, *32*, 143–157. [CrossRef]
2. Van der Salm, C.; Kros, J.; de Vries, W. Evaluation of different approaches to describe the sorption and desorption of phosphorus in soils on experimental data. *Sci. Total Environ.* **2016**, *571*, 292–306. [CrossRef]
3. Xu, X.; Qiu, H.; Zhou, X. The absorption, Translocation and Metabolism of Phosphorus of Plant. *J. Shandong Agric. Univ.* **2001**, *3*, 397–400.
4. Zhao, Q.; Zeng, D. Phosphorus cycling in terrestrial ecosystems and its controlling factors. *Acta Phytoecol. Sin.* **2005**, *1*, 153–163.
5. Miguel, M.A.; Postma, J.A.; Lynch, J.P. Phene Synergism between Root Hair Length and Basal Root Growth Angle for Phosphorus Acquisition. *Plant Physiol.* **2015**, *167*, 1430–1439. [CrossRef]
6. Yu, J.; Yin, D.; Wu, J.; Zhou, C.; Ma, X. A review of Adaptation Mechanism of Trees under low Phosphorus stress. *World For. Res.* **2007**, *30*, 18–23.
7. Gu, X. Effects of Planting Methods and Nitrogen Application Rate on Soil Environment and Yield of Winter Rapeseed. Ph.D. Thesis, Northwest Agriculture and Forestry University, Xianyang, China, 2018.
8. Cathcart, J.B. World phosphate reserves and resources. In *The Role of Phosphorus in Agriculture*; Khasawneh, F.E., Sample, E.C., Kamprath, E.J., Eds.; American Society of Agronomy: Madison, WI, USA, 1980; pp. 1–18.
9. Li, Y.; Xiao, Z. China's Forestland Soil Pollution, Degradation, Erosion Problems and Countermeasures. *For. Econ.* **2015**, *37*, 3–15.
10. Kale, R.R.; Anila, M.; Swamy, H.M.; Bhadana, V.P.; Rani, C.V.D.; Senguttuvel, P.; Sundaram, R.M. Morphological and molecular screening of rice germplasm lines for low soil P tolerance.Journal of plant biochemistry and biotechnology. *J. Plant Biochem. Biotechnol.* **2020**, *10*, 1–12.
11. Lambers, H.; Plaxton, W.C. Phosphorus:Back to the roots. *Annu. Plant Rev.* **2015**, *48*, 3–22.
12. Da Silva, A.; Bruno, I.P.; Franzini, V.I.; Marcante, N.C.; Benitiz, L.; Muraoka, T. Phosphorus uptake efficiency, root morphology and architecture in Brazilian wheat cultivars. *J. Radioanal. Nucl. Chem.* **2016**, *7*, 1055–1063. [CrossRef]
13. Yao, J.; Zhou, Z.; Chu, X. Effect of neighborhood competition on dry matter accumulation, nitrogen and phosphorus efficiency of three provenances of Schima superba in a heterogeneous nutrient environment. *Acta Ecol. Sin.* **2018**, *38*, 1780–1788.
14. He, X.; Qi, B.; Wang, M.; Sun, Y. Differences in Biomass and Phosphorus Nutrition of Oats with Different Phosphorus Efficiency under Low Phosphorus Stress. *Mol. Plant Breed.* **2019**, *38*, 1780–1788.
15. Xu, X.; Zhang, Y. Research progress on the root adaptation mechanism of plants under low phosphorus stress. *Jiangsu J. Agric. Sci.* **2018**, *34*, 1425–1429.
16. Liu, P.; Wu, A.; Wang, J. Study on Phosphorus Use Efficiency and Phosphorus Remobilization Characteristics of Four Different Sorghum Genotypes. *J. Shanxi Agric. Sci.* **2018**, *46*, 344–349.
17. Li, T.; Ye, D.; Zhang, X.; Guo, J. Research advances on response characteristics of plants to different forms of phosphorus. *J. Plant Nutr. Fertil.* **2017**, *23*, 1536–1546.
18. Wu, W.; Wang, P.; Chen, N. Root foraging ability for phosphorus of different Chinese fir family seedlings under low phosphorus supply. *J. Fujian Agric. For. Univ.* **2019**, *48*, 174–181.
19. Aziz, T.; Steffens, D.; Rahmatullah; Schubert, S. Variation in phosphorus efficiency among brassica cultivars II:Changes in root morphology and carboxylate exudation. *J. Plant Nutr.* **2011**, *34*, 2127–2138. [CrossRef]
20. Wang, Z.; Hua, J.; Yin, Y. An Integrated Transcriptome and Proteome Analysis Reveals Putative Regulators of Adventitious Root Formation in Taxodium 'Zhongshanshan'. *Int. J. Mol. Sci.* **2019**, *20*, 1225. [CrossRef]
21. Yu, C.; Xu, S.; Yin, Y. Transcriptome analysis of the Taxodium 'Zhongshanshan 405' roots in response to salinity stress. *Plant Physiol. Biochem.* **2016**, *100*, 156–165. [CrossRef]
22. Wang, Z.; Gu, C.; Xuan, L.; Hua, J.; Shi, Q.; Fan, W.; Yin, Y.; Yu, F. Identification of suitable reference genes in Taxodium 'Zhongshanshan' under abiotic stresses. *Trees* **2017**, *31*, 1519–1530. [CrossRef]
23. Qi, B.; Yang, Y.; Yin, Y.; Xu, M.; Li, H. De novo sequencing, assembly, and analysis of the Taxodium'Zhongshansa'roots and shoots transcriptome in response to short-term waterlogging. *BMC Plant Biol.* **2014**, *14*, 201. [CrossRef]

24. Jiang, H.; Zhang, L.; Zhang, Q. Effects of waterlogging stress on photosynthesis and fluorescence of Taxodium hybrid. *J. Fujian Agric. For. Univ.* **2019**, *48*, 447–452.
25. Shi, Q.; Yin, Y.; Wang, Z. Response in cuttings of Taxodium hybrid'Zhongshanshan'and their parents to drought and re-hydration. *Chin. J. Appl. Ecol.* **2016**, *27*, 3435–3443.
26. Yu, C.; Xu, J.; Lu, Z.; Yin, Y. A regional experiment of new cultivars of Taxodium mucronatum hybrid. *South China For. Sci.* **2015**, *43*, 27–30. [CrossRef]
27. Yin, Y.; Chen, Y.; Chaoguang, Y.U. Breeding and popularization of new hybrid varieties of Taxodium Rich. *Trees Sci. Technol. Achiev.* **2013**.
28. Xu, J.; Yin, Y.; Hua, J. Analysis on Growth Characteristics of Three Taxodium mucronatum Hybrids. *Mod. Agric. Sci. Technol.* **2015**, *24*, 144–146, 150.
29. Li, C.; Wang, J.; Chen, X. Experimental study on introduction of Taxodium hybrid'Zhongshanshan'. *J. Hebei For. Sci. Technol.* **2012**, *2*, 9–12.
30. Guo, J.; Shi, Q.; Xiong, Y.; Yin, Y.; Hua, J. Effects of salt-alkaline mixed stress on growth and photosynthetic characteristics of Taxodium hybrid 'Zhongshanshan 406'. *J. Nanjing For. Univ.* **2019**, *43*, 61–68.
31. Shen, J.; Mao, D. *Methods of Plant Nutrition Research*; China Agricultural University Press: Beijing, China, 2011.
32. Yang, S.; Zhang, H.X.; Yang, X.Y.; Chen, Q.X.; Wu, H.W. Differential Growth Performance of Elaeagnus angustifolia Provenances under NaCl Stress. *Sci. Silvae Sin.* **2015**, *51*, 51–58.
33. Chen, W.; OuYang, Z.; OuYang, S.; Li, Z.; He, Y. Effects of low phosphorus stress on growth and biomass of phyllostachys alba. *Huannan For. Sci. Technol.* **2020**, *47*, 25–30.
34. Xu, J.; Li, Q.; Wu, W.; Rashid Muhammad, H.U.; Ma, X.; Wu, P. Effects of vertical phosphor competition on root growth and biomass allocation of Chinese fir. *Acta Ecol. Sin.* **2019**, *39*, 2071–2081.
35. Yu, C.; Yin, Y.; Xu, J. Four hybrid varieties of Taxodium. *Sci. Silvae Sin.* **2011**, *47*, 181–182. (In Chinese)
36. Hua, J.; Yin, Y.; Zhou, D.; Yu, C.; Xu, J. Effects of soil water conditions on growth and physiology of Taxodium 'Zhongshanshan 406'. *J. Ecol. Rural Environ.* **2011**, *6*, 10.
37. Yuan, J. Study on Adaptive Mechanism of Camellia Oleifera to Low-Phosphorus Environment. Ph.D. Thesis, Beijing Forestry University, Beijing, China, 2013.
38. Li, B.; Xiao, K.; Li, Y. The genotypic differences in photosynthetic characteristics of wheat to low- phosphorous stress. *J. Agric. Univ. Hebei.* **2002**, *1*, 5–9.
39. Haynes, R.J. Active ion uptake and maintenance of cationanion balance: A critical examination of their role in regulating rhizosphere pH. *Plant Soil* **1990**, *126*, 247–264. [CrossRef]
40. Jones, D.L.; Darrah, P.R. Role of root derived organic acids in the mobilization of nutrients from the rhizosphere. *Plant Soil* **1994**, *166*, 247–257. [CrossRef]
41. Hinsinger, P. Bioavailability of soil inorganic P in the rhizosphere as affected by root-induced chemical changes: A review. *Plant Soil* **2001**, *237*, 173–195. [CrossRef]

MDPI

Article

Responses of Swamp Cypress (*Taxodium distichum*) and Chinese Willow (*Salix matsudana*) Roots to Periodic Submergence in Mega-Reservoir: Changes in Organic Acid Concentration

Xinrui He [1], Ting Wang [1,2], Kejun Wu [1,3], Peng Wang [1], Yuancai Qi [1], Muhammad Arif [1,4] and Hong Wei [1,*]

[1] Key Laboratory of Eco-Environments in Three Gorges Reservoir Region (Ministry of Education), Chongqing Key Laboratory of Plant Ecology and Resources Research in Three Gorges Reservoir Region, School of Life Sciences, Southwest University, TianSheng Road, Chongqing 400715, China; hxr2018@email.swu.edu.cn (X.H.); wangtingqy@gmail.com (T.W.); wkj123456@email.swu.edu.cn (K.W.); wp97206@email.swu.edu.cn (P.W.); qyc201520@email.swu.edu.cn (Y.Q.); muhammadarif@swu.edu.cn (M.A.)

[2] Dongguan Hanlin Experimental School, Dongguan 523000, China

[3] Sichuan Academy of Forestry, Chengdu 610081, China

[4] Punjab Forest Department, Government of Punjab, Lahore 54000, Pakistan

* Correspondence: weihong@swu.edu.cn; Tel.: +86-23-68-252-365

Citation: He, X.; Wang, T.; Wu, K.; Wang, P.; Qi, Y.; Arif, M.; Wei, H. Responses of Swamp Cypress (*Taxodium distichum*) and Chinese Willow (*Salix matsudana*) Roots to Periodic Submergence in Mega-Reservoir: Changes in Organic Acid Concentration. *Forests* **2021**, *12*, 203. https://doi.org/10.3390/f12020203

Academic Editor: Antonio Montagnoli

Received: 3 December 2020
Accepted: 6 February 2021
Published: 10 February 2021

Publisher's Note: MDPI stays neutral with regard to jurisdictional claims in published maps and institutional affiliations.

Abstract: Organic acids are critical as secondary metabolites for plant adaption in a stressful situation. Oxalic acid, tartaric acid, and malic acid can improve plant tolerance under waterlogged conditions. Two prominent woody species (*Taxodium distichum*-Swamp cypress and *Salix matsudana*-Chinese willow) have been experiencing long-term winter submergence and summer drought in the Three Gorges Reservoir. The objectives of the present study were to explore the responses of the roots of two woody species during flooding as reflected by root tissue concentrations of organic acids. Potted sample plants were randomly divided into three treatment groups: control, moderate submergence, and deep submergence. The concentrations of oxalic acid, tartaric acid, and malic acid in the main root and lateral roots of the two species were determined at four stages. The results showed that *T. distichum* and *S. matsudana* adapted well to the water regimes of the reservoir, with a survival rate of 100% during the experiment period. After experiencing a cycle of submergence and emergence, the height and base diameter of the two species showed increasing trends. Changes in base diameter showed insignificant differences between submergence treatments, and only height was significant under deep submergence. The concentrations of three organic acids in the roots of two species were influenced by winter submergence. After emergence in spring, two species could adjust their organic acid metabolisms to the normal level. Among three organic acids, tartaric acid showed the most sensitive response to water submergence, which deserved more studies in the future. The exotic species, *T. distichum*, had a more stable metabolism of organic acids to winter flooding. However, the native species, *S. matsudana*, responded more actively to long-term winter flooding. Both species can be considered in vegetation restoration, but it needs more observations for planting around 165 m above sea level, where winter submergence is more than 200 days.

Keywords: hydro-fluctuation zone; Three Gorges Dam Reservoir; winter submergence; *Taxodium distichum*; *Salix matsudana*; organic acids

1. Introduction

Reservoirs have played an essential role in human development throughout the world for thousands of years by providing water and controlling flooding, as well as generating power and electricity [1,2]. The Three Gorges Dam Reservoir is one of the largest reservoirs globally, located in the Yangtze River upper reaches in China [3]. Annually, the water level in the reservoir fluctuates by 30 m from 145 m to 175 m above sea level (ASL), the hydro-fluctuation zone is saturated in winter and dry in summer [4–6]. Therefore, many native plants sensitive to winter submergence are dying out gradually [7], which leads to negative environmental effects such as soil degradation and erosion, geology disasters, and biodiversity loss [2,8,9]. Conducting vegetation restoration in the hydro-fluctuation zone is important to improve the ecological quality and ecosystem services. In recent years, many studies have proposed that reforestation in the zone between 165 m and 175 m ASL may be more effective and sustainable [4,10]. Understanding the physiological and ecological mechanisms of flood-tolerant plants would be the first step. Such knowledge could be useful for the restoration in the riparian of the Three Gorges Reservoir.

Plants are exposed to diverse stress conditions throughout their life, including biotic and abiotic stresses [11]. Abiotic challenges are metal-toxicity, drought stress, flooding-stress, salt-stress, high or low temperature-stress, nutrient-deprivation, and light-stress [12–14]. Among them, water environmental stresses, including drought and flooding, usually negatively affected the growth of plants [13,15,16]. The diffusion rate of gasses such as oxygen and carbon dioxide under submergence is approximately one-ten-thousandth of that in the air [17], which could disturb the photosynthetic and aerobic respiration activity and then restrict the growth of plants [18,19]. Flooding results in the increase of toxic substances such as Fe^{2+}, Mn^{2+} in the soil, which can be absorbed by the plant; in addition, ethanol remaining in tissues can be converted into acetaldehyde when re-aeration after a period of O_2 deprivation [20]. Thus, excessive accumulation of Fe^{2+}, Mn^{2+}, and ethanol in the hypoxic state results in increased plant pathology [21,22]. Moreover, plants can suffer severe oxidative stress due to high levels of reactive oxygen species (ROS) resulting from metabolic changes following the alternation of flood and drought [23]. Thus, plant growth and survival in the water-level fluctuation zone of Three Gorges Reservoir can be challenging.

Numerous investigations have focused on the water tolerance mechanisms of plants, such as the anatomy and physiological changes under waterlogging submergence conditions [24], rapid formation of adventitious roots and aerenchyma under flooding [25,26], changes in carbohydrate content, and antioxidant enzymatic activities [27,28], and photosynthetic physiological adaption to submergence [24,29]. Many studies have also focused on the physiology during the recovery period from submergence [29–32]. It is essential to examine the adaptation processes of plants exposed to long-term periodic submergence and drought. However, the physiological characteristics of plants to dynamic water levels of reservoir still remain unclear, as does whether flooding-tolerant plants respond similarly to submergence.

The root system is the main organ by which plants absorb water and nutrients, but it is susceptible to environmental changes. Low molecular weight organic acids are secondary metabolites produced in plants under environmental stress conditions that can remove free radicals and relieve toxicity [33–35]. It also strengthens disease resistance, activates insoluble soil phosphorus, chelates heavy metal in contaminated areas, and enhances antioxidant enzyme activity [36,37]. Oxalic acid, tartaric acid, and malic acid, as critical organic acids, can improve the tolerance of waterlogged plants to water stress [24,38–40]. How do these organic acids respond to the hydro-fluctuation in the Three Gorges Reservoir to help flooding-tolerant plants survive?

In the preliminary study, a batch of waterlogged plants screened out has been applied to reforest in the hydro-fluctuation zone. It was found that the woody plant species *Taxodium distichum* and *Salix matsudana* can grow well after repeated annual flooding [29,31,40,41]. *T. distichum*, native to Southeastern North America, has a specialized respiratory root structure for underwater conditions and was introduced into the hydro-fluctuation zone due to its flooding-resistant characteristics [42]. *S. matsudana* is a native tree in the Three Gorges Reservoir region. They both grow well in the hydro-fluctuation zone of the Three Gorges Reservoir, and both are considered candidates for the incorporation of reforestation. The objectives of this study were to explore the responses of the roots of two woody species during flooding conditions. The concentration of organic acids can reflect the reactions to dynamic water stress caused by submergence in winter and drought in summer. This research can recognize the physiological and ecological adaptation mechanisms of *T. distichum* and *S. matsudana* as reforestation species in the hydro-fluctuation zone of Three Gorges Reservoir. We hypothesized that: (1) Both species have strong waterlogging tolerance, but high-intensity waterlogging would inhibit plant growth to some degree. (2) Oxalic acid, tartaric acid, and malic acid in the two suitable plant roots will show increasing responses to a certain degree when suffering from the dynamic variation of soil moisture in the water-fluctuation zone. (3) The growth and adaptability of two woody plants are related to their characteristics of organic acid concentrations. The findings from the water rhythms in situ will enrich researcher knowledge about the adaptive tree species and benefit the reforestation practices in the Three Gorges Reservoir region.

2. Materials and Methods

2.1. Study Region

The experimental Plot (107°32′–108°14′ E, 30°03′–30°35′ N) is located along the Ruxi River, Shibaozhai, Gonghe Village, Zhong County, Chongqing Municipality, China. Ruxi River is a tributary of the Yangtze River, with a subtropical, southeastern monsoonal mountain climate. The region experiences an annual accumulative temperature of 5787 °C for days ≥10 °C, an annual average temperature of 18.2 °C, and highest and lowest temperatures of 40 °C and 0 °C, respectively. The highest temperatures mainly occur from July to September. There is a frost-free period of 341 days, total solar radiation of 83.7×4.18 kJ/cm^{-2}, and an annual sunshine ratio of 29%. The mean annual precipitation is 1200 mm, and most rainfalls are in May and June. Plants suffer drought due to high air temperature and relatively fewer rainfalls in July and August [43,44].

The plants are experiencing annual cycles of periodic submergence and drought growing in the hydro-fluctuation zone of the Three Gorges Dam (Figure 1). Water level gradually rises in the fluctuation zone from September to October. It attains the highest altitude (175 m ASL) from November through to January, then gradually decreases, reaching its lowest level at 145 m ASL from June to September.

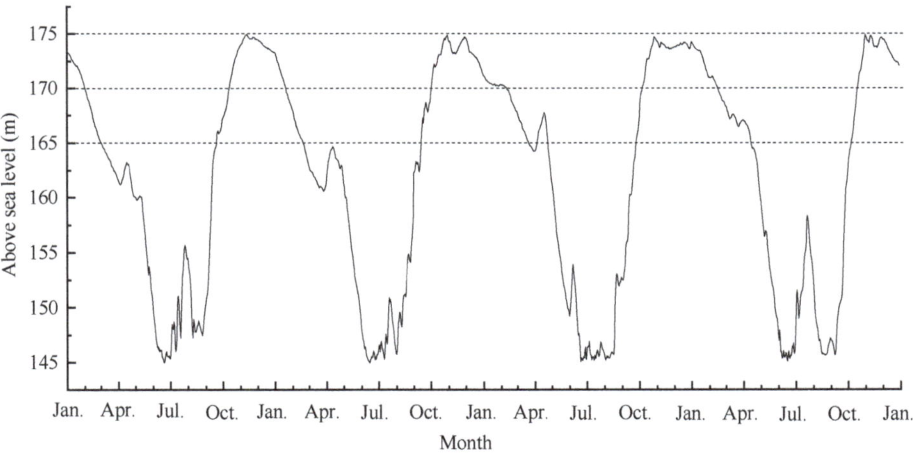

Figure 1. The water-level changes in the hydro-fluctuation zone of the Three Gorges Reservoir from January 2013 to January 2017.

2.2. Experiment Design

Two-year-old potted seedlings of *T. distichum* and *S. matsudana* were cultivated in a greenhouse in Shibaozhai, Zhong County, Chongqing, China, close to the study area, with one plant per pot filled with 12.5 kg of purple soil from the water-level fluctuation zone of the Three Gorges Reservoir.

On 15 September 2015 (T0), for each species, 36 seedlings of uniform growth were randomly divided into three groups and placed in situ at an altitude of 175 m to 165 m ASL. The mean plant height and base diameter at T0 for *T. distichum* were 69.1 cm and 8.38 mm, compared with 136.25 cm and 17.82 mm for *S. matsudana*. The three treatment groups were shallow submergence, serving as control (shallow submergence, abbreviated as SS, 175m ASL), periodic moderate submergence (MS, 170 m ASL), and deep submergence (DS, 165 m ASL). The maximum submergence depth and submergence days in each treatment were shown in Table 1. At the initial stage of water withdrawal (T1, 16 April 2016), the stage of the recovery period (T2, 15 July 2016), and the beginning of the next submergence period (T3, 15 September 2016), 4 plants of *T. distichum* and *S. matsudana* were sampled, respectively, in each treatment for further measurements.

Table 1. Water level changes in situ following treatments at different elevations of the Three Gorges Reservoir.

Elevation (m)	Treatment Group	Maximum Submergence Depth (m)	Experiment Duration (Days)		
			T0–T1	T1–T2	T2–T3
175	SS	0	0	295	70
170	MS	5	135	160	70
165	DS	10	205	90	70

SS, control; MS, moderate submergence; DS, deep submergence; T0, 15 September 2015; T1, 16 April 2016; T2, 15 July 2016; T3, 15 September 2016.

2.3. Sample Analysis

Before collecting each sample, plant height and base diameter of two species were measured using a height measuring rod and vernier calipers, respectively.

Primary roots were separated from the laterals root of each plant when collecting samples and immediately transported to the laboratory in an icebox. They were placed in an oven at 110 °C for 15 min after washing with deionized water, then adjusted to 80 °C to dry until constant weight [45,46], and finally ground into a powder then sifted through a 1 mm sieve.

A 0.1 g sample of root powder was placed in a 10 mL centrifuge tube with 5 mL of ultrapure water and ultrasonicated for 1 h, then cooled to room temperature. After centrifugation at 8000 rpm for 10 min, the supernatant was filtered through a 0.45 μm syringe filter (Millipore, Danvers, MA, USA) to determine the concentrations of oxalic acid, tartaric acid, and malic acid [47].

A high-performance liquid chromatography method was used to determine the organic acid concentration in dried roots by the Sepax Sapphire C18 column (4.6 mm × 250 mm, 5 μm). The mobile phase aqueous and organic components were 95% 20 mmol·L^{-1} KH_2PO_4 buffer (pH 2.5, adjusted with phosphoric acid) and 5% methanol. The HPLC instrument was coupled to an Agilent 1100 diode array multi-wavelength detector. The mobile phase flow rate was 0.9 mL·min^{-1}, the detection wavelength was 210 nm, the column temperature was 30 °C, and the injection volume was 20 μL [47,48]. Organic acids in main and lateral roots were measured in milligrams of organic acid per gram of the mains root and lateral roots dry matter (mg.g^{-1}, DW).

2.4. Statistical Analysis

SPSS 22.0 was used for statistical analyses. Paired sample *t*-tests were used to analyze the differences of growth between different experimental periods for the same treatment (Table 2). Tukey's tests were applied to determine the significance of differences between different submerging treatment groups (Table 2). Repeated measures ANOVA was conducted to assess the effects of submergence, experimental period, and their interaction on the concentrations of organic acids for *T. distichum* and *S. matsudana* (Table 3). Repeated measurement analysis of variance and Bonferroni method were carried out to analyze the differences of organic acids among different experimental periods for the same plant [49] (Figures 2 and 3). One-way ANOVA and Tukey's tests were carried out to analyze the differences of organic acids among different treatment groups for the same plant (Figure 4). The p-value for statistical tests was 0.05.

Table 2. Paired sample t-tests and Tukey's tests for growth of *Taxodium distichum* and *Salix matsudana*.

Indices	Treatment Group	*T. distichum*		*S. matsudana*	
		15 September 2015	15 September 2016	15 September 2015	15 September 2016
Height(cm)	SS	69.10 ± 0.59 Aa	86.87 ± 0.81 Ba	136.25 ± 0.88 Aa	176.75 ± 0.99 Ba
	MS	69.10 ± 0.59 Aa	80.25 ± 0.42 Ba	136.25 ± 0.88 Aa	171.50 ± 1.25 Ba
	DS	69.10 ± 0.59 Aa	72.30 ± 0.87 Ab	136.25 ± 0.88 Aa	146.50 ± 0.80 Ab
Base diameter(mm)	SS	8.38 ± 0.54 Aa	11.00 ± 0.64 Ba	17.82 ± 0.86 Aa	20.00 ± 0.55 Aa
	MS	8.38 ± 0.54 Aa	9.37 ± 0.59 Aa	17.82 ± 0.86 Aa	19.02 ± 0.83 Aa
	DS	8.38 ± 0.54 Aa	9.13 ± 0.35 Aa	17.82 ± 0.86 Aa	18.55 ± 0.77 Aa

Data showed means ± standard error (SE; n = 4). SS, control; MS, moderate submergence; DS, deep submergence. Values with different capital letters within the same line indicate significant differences between 15 September 2015, and 15 September 2016 ($p < 0.05$). Values with different small letters in the same column are significantly different among different submergence treatments ($p < 0.05$).

Table 3. Repeated measures ANOVA for organic acid concentrations in roots of *T. distichum* and *S. matsudana*.

Variable	Treatment			Experimental Period			Treatment × Experimental Period		
	df	*F*		*df*	*F*		*df*	*F*	
		T. distichum	*S. matsudana*		*T. distichum*	*S. matsudana*		*T. distichum*	*S. matsudana*
Oxalic acid in main roots	2	0.103 ^{ns}	0.414 ^{ns}	3	56.075 *	16.995 *	6	1.238 ^{ns}	2.078 ^{ns}
Oxalic acid in lateral roots	2	0.034 ^{ns}	3.463 *	3	1.242 ^{ns}	12.034 *	6	1.157 ^{ns}	5.202 *
Tartaric acid in main roots	2	4.856 *	14.412 *	3	17.263 *	8.404 *	6	5.347 *	1.570 ^{ns}
Tartaric acid in lateral roots	2	13.672 *	10.243 *	3	14.659 *	4.282 *	6	1.179 ^{ns}	1.153 ^{ns}
Malic acid in main roots	2	17.779 *	2.374 ^{ns}	3	14.847 *	1.388 ^{ns}	6	6.564 *	1.343 ^{ns}
Malic acid in lateral roots	2	0.232 ^{ns}	0.599 ^{ns}	3	4.588 *	15.792 *	6	0.232 ^{ns}	1.093 ^{ns}

^{ns} $p > 0.05$; * $p < 0.05$.

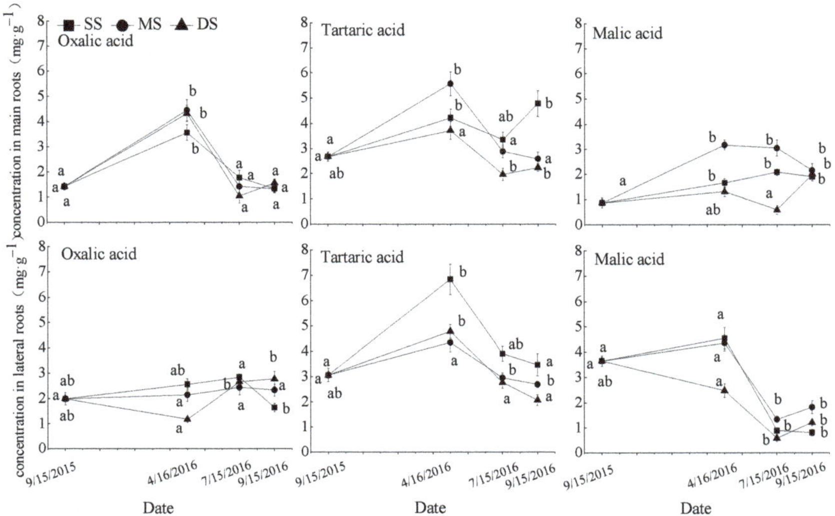

Figure 2. Repeated measurement analysis of variance and Bonferroni method for oxalic acid, tartaric acid, and malic acid concentrations in roots of *T. distichum*. Data are shown as means ± SE (n = 4). SS, control; MS, moderate submergence; DS, deep submergence. Values with different letters indicate significant differences among different sampling times for the same treatment ($p < 0.05$).

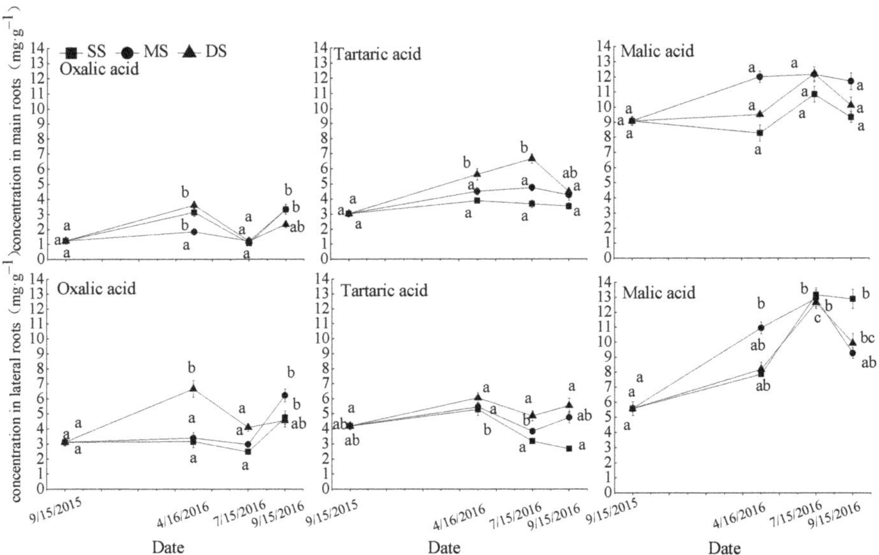

Figure 3. Repeated measurement analysis of variance and Bonferroni method for oxalic acid, tartaric acid, and malic acid concentrations in roots of in *S. matsudana*. Data are shown as means ± SE (n = 4). SS, control; MS, moderate submergence; DS, deep submergence. Values with different letters indicate significant differences among different sampling times for the same treatment ($p < 0.05$).

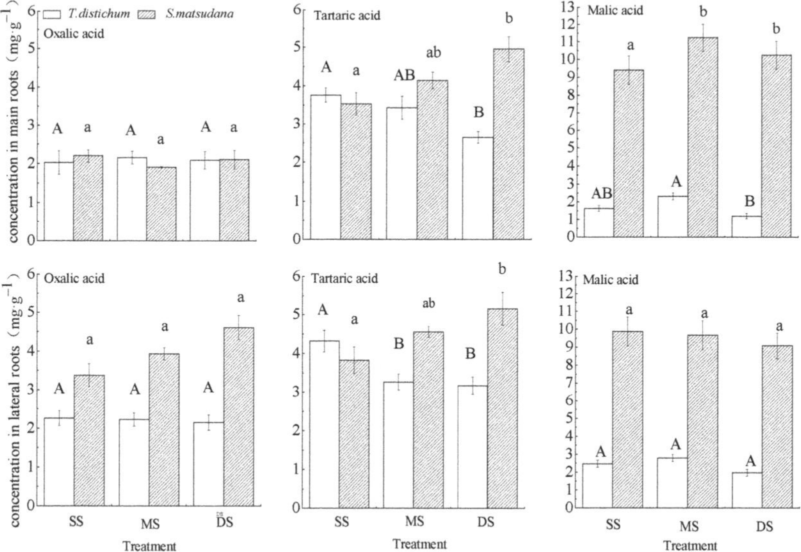

Figure 4. One-way ANOVA and Tukey's tests to compare organic acid concentrations in the roots of *T. distichum* and *S. matsudana* under different submergence treatments. Data are shown as means ± SE (n = 4). SS, control; MS, moderate submergence; DS, deep submergence. Values with different letters are significantly different among different submergence treatments. Capital letters represent *T. distichum* seedlings, and small letters indicate *S. matsudana* seedlings.

3. Results

3.1. Growth of T. distichum and S. matsudana Plants

The survival rate was 100% throughout the experiment for both species. The heights and base diameters of *T. distichum* and *S. matsudana* plants were increased after one year of growth (Table 2). Compared to the initial values at T0 (15 September 2015), the plant height of two species in the SS and MS groups and the base diameter of *T. distichum* in the SS group at T3 (15 September 2016) were significantly higher. Winter submergence significantly inhibited the growth height of plants in the DS groups. In contrast, the base diameter of *T. distichum* and *S. matsudana* plants showed fewer differences among different treatment groups.

3.2. Organic Acid Concentrations in T. distichum Roots

Table 3 shows the repeated measures of ANOVA results for organic acid concentrations in the roots of *T. distichum* following different water treatments, experimental periods, and their interaction. The experimental period had the most significant effect on the concentrations of organic acids in the root system. Different water treatments significantly affected tartaric acid in the main and lateral root and malic acid in the main root. In addition, tartaric acid and malic levels in the main root were significantly affected by treatment and experimental period interaction.

The response of the oxalic acid, tartaric acid, and malic acid concentrations in *T. distichum* roots to soil moisture was shown in Figure 2. The concentrations of three organic acids in the main and lateral roots displayed similar responses to the changes in soil moisture in the fluctuating water zone. Most of them increased during the flooding stress period (T0–T1), then decreased during the recovery growth period (T1–T2), and were no significant changes under mild drought stress (T2–T3). The concentrations of three organic acids in roots were influenced by winter submergence. The concentrations of tartaric acid in lateral roots in the MS and DS groups and malic acid in lateral roots in the DS group were significantly decreased compared with that in SS group. After emergence in spring, three organic acid concentrations had no differences among the three water treatments.

3.3. Organic Acid Concentrations in S. matsudana Roots

Different flooding treatments in the water fluctuation zone significantly affected oxalic acid concentration in the lateral roots and the tartaric acid concentrations in the main and lateral roots of *S. matsudana* (Table 3). Except for malic acid in the main roots, oxalic acid and tartaric acid in both the main and lateral roots, and malic acid in the lateral roots of *S. matsudana* were significantly affected by different experimental periods. The interaction between treatment and experimental periods had little effect on organic acids in the roots of *S. matsudana*. Only the oxalic acid concentration in lateral roots was significantly affected.

The response of the oxalic acid, tartaric acid, and malic acid concentrations in *S. matsudana* roots to soil moisture was shown in Figure 3. The concentrations of oxalic acid in the main and lateral roots and tartaric acid in the lateral roots were increased during the flooding stress period (T0–T1), then decreased during the recovery growth period (T1–T2), and were no significant changes under mild drought stress (T2–T3). The concentrations of malic acid in the main and lateral roots and tartaric acid in the main roots increased during the flooding stress period (T0–T1) and the recovery growth period (T1–T2), and then decreased under mild drought stress (T2–T3). Under water submergence in winter, the concentration of oxalic acid in the lateral roots and tartaric acid in the main roots were significantly increased. All acid concentrations in each treatment showed no significant difference after emergence from winter flooding.

3.4. Comparison between T. distichum and S. matsudana

Figure 4 showed the difference in mean concentrations of three organic acids in the whole experimental period between two species. The concentrations of organic acids in the roots of *S. matsudana* were higher than those of *T. distichum* generally.

The effects of different treatments on the concentrations of oxalic acid in the main and lateral roots of *T. distichum* and *S. matsudana* were similar. Under flooding stress, there were no significant differences in oxalic acid concentrations in the main and lateral roots of *T. distichum* and *S. matsudana* among the three treatment groups. The tartaric acid concentrations in the main and lateral roots of *T. distichum* gradually decreased with increasing flooding intensity, but it was the opposite trend for *S. matsudana*. The malic acid concentrations in the main roots of *S. matsudana* in MS and DS groups were significantly higher than those in the SS group. No significant differences in the concentrations of malic acid in lateral roots of *T. distichum* and *S. matsudana* were showed. The malic acid concentrations in the main and lateral roots were more sensitive to water flooding in *S. matsudana* than *T. distichum*.

4. Discussion

At present, adaptive woody plants such as *T. distichum* and *S. matsudana* were screened by simulated flooding tests and in situ planting analysis [23]. This study has shown that both *T. distichum* and *S. matsudana* adapt well to soil regimes in the water fluctuation zone of Three Gorges Reservoir. The seedlings of two species survived and grown well after winter submergence, although the flooding more than 200 days at 165 m ASL still inhibited the growth height of plants to a certain degree (Table 2). The plants planted around 165 m ASL should be paying more attention to their growth and survival after years of periodical winter flooding. The growth of *T. distichum* and *S. matsudana* at 170 m ASL recovered significantly after the emergence and showed no significant differences with plants at 175 m ASL.

Flooding is often accompanied by secondary stresses such as low temperature, low light, and low dissolved oxygen. The photosynthesis of plants is limited under flooding. Studies have shown that the survival of plants under flooding stress and the restoration of growth after flooding are closely related to the regulation of certain metabolic pathways [18]. Organic acids are secondary metabolites related to adaptation that play important roles in defenses against adverse conditions. The responses of oxalic acid, tartaric acid, and malic acid in roots of *T. distichum* and *S. matsudana* were different to winter flooding in the hydro-fluctuation zone. However, two species could adjust their organic acid metabolisms to the normal level after emergence in spring.

Under winter flooding stress, relative to the control group, the tartaric acid concentrations in the main and lateral roots and the malic acid concentration in the main roots of *T. distichum* in the DS group decreased significantly. However, the concentrations of tartaric acid in the main and lateral roots of *S. matsudana* in the DS group increased significantly, and the malic acid concentration in the main roots of the flooded group was increased significantly, while the other organic acids were not altered significantly (Figure 4). Thus, tartaric acid and malic acid responses in *S. matsudana* roots were more active than those in *T. distichum* roots.

Oxalic acid is one of the most important secondary metabolites in plants under adverse stress conditions, it can stimulate peroxidase activity and improves the antioxidant capacity [33]. Herein, the concentrations of oxalic acid in the roots of *T. distichum* and *S. matsudana* were greatly affected by the experimental period but not significantly affected by different flooding treatments. The concentration of oxalic acid in the main roots of *T. distichum* was significantly increased at the initial stage of water withdrawal (T1), and gradually decreased during the normal growth period; no significant changes showed under mild drought stress (Figure 2). The concentrations of oxalic acid in the main and lateral roots of *S. matsudana* increased gradually under water flooding stress, and remained high during the initial stages of plant reoxygenation, then decreased during the normal growth period, until the drought period (Figure 4), consistent with the simulation results [40]. Metabolic imbalance of active oxygen in flooding and the early stage of reoxygenation can result in severe oxidative stress. Thus, increased oxalic acid concentrations in the main roots of *T. distichum* and the main and lateral roots of *S. matsudana* may be associated with increased antioxidant capacity. During the recovery stage, reactive oxygen species production and elimination gradually achieve a balance; the oxalic acid concentration of roots may provide energy for plant growth through catabolic pathways, resulting in a decrease in oxalic acid concentration during recovery growth after emergence in spring. Throughout the whole experiment, the concentration of oxalic acid in the main roots of *T. distichum* was more sensitive than that in the lateral roots, while the concentration of oxalic acid in the lateral roots of *S. matsudana* was more sensitive than that in the main roots, but the roots of both plants responded positively to flooding in the water fluctuation zone. Relevant studies have pointed out that lateral root plays a more important role in responding to abiotic stress for plants [50], and adaptive plants can make full use of lateral root metabolism regulation to enhance the ability of water flooding tolerance [23].

The concentrations of tartaric acid in the main and lateral roots of *S. matsudana* was ordered by DS group > MS group > SS group. Some studies have previously reported that organic acids not only function as important intermediates in the energy balance and flow in plants but also participate in plant adaption to abiotic stress [51,52]. The DS group maintained a higher root tartaric acid concentration than the MS group. Therefore, we speculated that tartaric acid in the *S. matsudana* root system could respond positively to flooding stress in the water fluctuation zone [39]. The concentration of tartaric acid in *S. matsudana* differed between main and lateral roots, indicating different response strategies. Studies have shown that the change of organic acid concentration is closely related to the adaptation mechanism to environmental stress [52].In this study, changes in tartaric acid concentration in *S. matsudana* lateral roots over time were similar to those of oxalic acid in roots. The concentrations of oxalic acid and tartaric acid in roots are controlled by synthesis, decomposition, transport, and secretion [53], indicating that tartaric acid may perform similar physiological functions to oxalic acid upon changes in soil moisture in the water fluctuation zone. The synthesis of plant oxalic acid is mainly through the photorespiration glycolic acid pathway [54]. However, there is limited research work available on tartaric acid concentration mechanism under water-stressed conditions, and it is very important to be carried out further studies in this field.

5. Conclusions

The exotic species *T. distichum* and native species *S. matsudana* are two adaptive woody plants for reforestation in the hydro-fluctuation zone of the Three Gorges Reservoir. The two species can grow very well between 170 m to 175 m ASL in the hydro-fluctuation zone of Three Gorges Reservoir. It needs more observations for plants planted around 165 m ASL, where the duration of winter submergence is more than 200 days.

The concentrations of three organic acids in the roots of two species were influenced by winter submergence. After emergence in spring, two species could adjust their organic acid metabolisms to the normal level. Among three organic acids, tartaric acid was the most sensitive response to water submergence, which deserved more studies in the future. The exotic species, *T. distichum*, had more stable metabolism of organic acids to winter flooding. The native species, *S. matsudana*, responded more actively to long-term winter flooding. Two species showed the different organic acid regulation mechanisms facing water regimes, and both species can be applied in vegetation restoration.

Author Contributions: Funding acquisition, H.W.; investigation, X.H.; methodology, X.H., T.W., and K.W.; project administration, H.W.; resources, H.W.; software, X.H., K.W., and Y.Q; supervision, H.W. and M.A.; validation, X.H., H.W., and M.A.; visualization, X.H., P.W., and Y.Q.; writing—original draft preparation, X.H.; writing—review and editing, M.A. and H.W. All authors have read and agreed to the published version of the manuscript.

Funding: This research was jointly supported by the National Key Research and Development Program of China (2017YFC0505305), the Program for the Follow-on Work of the Three Gorges: Ecological and Biological Diversity Conservation in the Reservoir Region (5000002013BB5200002), and the Key Research Projects in Forestry of Chongqing (Yu Lin Ke Yan 2015-6).

Data Availability Statement: All data generated or analysed during this study are included in this published article.

Acknowledgments: The authors would like to acknowledge support from the Key Laboratory of Eco-environments in the Three Gorges Reservoir Region (Ministry of Education), the Chongqing Key Laboratory of Plant Ecology and Resources Research in Three Gorges Reservoir Region. We are grateful to Muhammad Tahir, Research Fellow at the Clausthal University of Technology, Germany, and Pinky Sanelisiwe Mzondi, Researcher at the Southwest University, China, for improving the language of this article.

Conflicts of Interest: The authors declare no conflict of interest.

References

1. Wu, J.G.; Huang, J.H.; Han, X.G.; Gao, X.M.; He, F.L.; Jiang, M.X.; Jiang, Z.G.; Primack, R.B.; Shen, Z.H. The Three Gorges Dam: An ecological perspective. *Front. Ecol. Environ.* **2004**, *2*, 241–248. [CrossRef]
2. New, T.; Xie, Z. Impacts of large dams on riparian vegetation: Applying global experience to the case of China's Three Gorges Dam. *Biodivers Conserv.* **2008**, *17*, 3149–3163. [CrossRef]
3. Zhu, N.N.; Qin, A.L.; Guo, Q.S.; Zhu, L.; Xu, G.X.; Pei, S.X. Spatial heterogeneity of plant community in Zigui and Wushan typical hydro-fluctuation belt of Three Gorges Reservoir areas. *For. Res.* **2015**, *28*, 109–115. [CrossRef]
4. Fan, D.Y.; Xiong, G.M.; Zhang, A.Y.; Li, X.; Xie, Z.G.; Li, Z.J. Effect of water-lever regulation on species selection for ecological restoration practice in the water-level fluctuation zone of Three Gorges Reservoir. *Chin. J. Plant Ecol.* **2015**, *39*, 416–432. [CrossRef]
5. Ye, C.; Li, S.Y.; Zhang, Y.R.; Zhang, Q.F. Assessing soil heavy metal pollution in the water-level-fluctuation zone of the Three Gorges Reservoir, China. *J. Hazard. Mater.* **2011**, *191*, 366–372. [CrossRef]
6. Wang, Q.; Yuan, X.Z.; Willison, J.H.M.; Zhang, Y.W.; Liu, H. Diversity and above-ground biomass patterns of vascular flora induced by flooding in the drawdown area of China's Three Gorges Reservoir. *PLoS ONE* **2014**, *9*, e100889. [CrossRef] [PubMed]
7. Garssen, A.G.; Baattrup-Pedersen, A.; Voesenek, L.A.C.J.; Verhoeven, J.T.A.; Soons, M.B. Riparian plant community responses to increased flooding: A meta-analysis. *Glob. Chang. Biol.* **2015**, *21*, 2881–2890. [CrossRef]
8. Kaczmarek, H.; Mazaeva, O.A.; Kozyreva, E.A.; Babicheva, V.A.; Tyszkowski, S.; Rybchenko, A.A.; Brykała, D.; Bartczak, A.; Słowiński, M. Impact of large water level fluctuations on geomorphological processes and their interactions in the shore zone of a dam reservoir. *J. Great Lakes Res.* **2016**, *42*, 926–941. [CrossRef]
9. Bao, Y.H.; Gao, P.; He, X.B. The water-level fluctuation zone of Three Gorges Reservoir—A unique geomorphological unit. *Earth Sci. Rev.* **2015**, *150*, 14–24. [CrossRef]
10. Li, C.X.; Zhong, Z.C. Influences of mimic soil water change on the contents of malic acid and shikimic acid and root-biomasses of *Taxodium distichum* seedlings in the hydro-fluctuation belt of the Three Gorges reservoir region. *Acta Ecol. Sin.* **2007**, *27*, 4394–4402. [CrossRef]
11. Cao, Y.Y.; Yang, M.T.; Ma, W.X.; Sun, Y.J.; Chen, G.Y. Overexpression of SSB_{Xoc}, a Single-Stranded DNA-Binding Protein from *Xanthomonas oryzae pv. oryzicola*, Enhances Plant Growth and Disease and Salt Stress Tolerance in Transgenic *Nicotiana benthamiana*. *Front. Plant Sci.* **2018**, *9*, 953. [CrossRef] [PubMed]
12. Dalal, V.K.; Tripathy, B.C. Water-stress induced downsizing of light-harvesting antenna complex protects developing rice seedlings from photo-oxidative damage. *Sci. Rep.* **2018**, *8*, 5955. [CrossRef]
13. Bhusal, N.; Kim, H.S.; Han, S.G.; Yoon, T.M. Impact of drought stress on photosynthetic response, leaf water potential, and stem sap flow in two cultivars of bi-leader apple trees (*Malus* × *domestica* Borkh.). *Sci. Hortic.* **2019**, *246*, 535–543. [CrossRef]
14. Viehweger, K. How plants cope with heavy metals. *Bot. Stud.* **2014**, *55*. [CrossRef] [PubMed]
15. Bangar, P.; Chaudhury, A.; Tiwari, B.; Kumar, S.; Kumari, R.; Bhat, K.V. Morphophysiological and biochemical response of mungbean [*Vigna radiata* (L.) Wilczek] varieties at different developmental stages under drought stress. *Turk. J. Biol.* **2019**, *43*, 58–69. [CrossRef]
16. Peng, Y.J.; Zhou, Z.X.; Zhang, Z.; Yu, X.L.; Zhang, X.Y.; Du, K.B. Molecular and physiological responses in roots of two full-sib poplars uncover mechanisms that contribute to differences in partial submergence tolerance. *Sci. Rep.* **2018**, *8*. [CrossRef]
17. Colmer, T.D. Long-distance transport of gases in plants: A perspective on internal aeration and radial oxygen loss from roots. *Plant Cell Environ.* **2003**, *26*, 17–36. [CrossRef]
18. Fukao, T.; Bailey-Serres, J. Plant responses to hypoxia—is survival a balancing act? *Trends Plant Sci.* **2004**, *9*, 449–456. [CrossRef]
19. Voesenek, L.A.C.J.; Colmer, T.D.; Pierik, R.; Millenaar, F.F.; Peeters, A.J.M. How plants cope with complete submergence. *New Phytol.* **2006**, *170*, 213–226. [CrossRef]
20. Bailey-Serres, J.; Voesenek, L.A.C.J. Flooding stress: Acclimations and genetic diversity. *Annl. Rev. Plant Biol.* **2008**, *59*, 313. [CrossRef]
21. Colmer, T.D.; Voesenek, L.A.C.J. Flooding tolerance: Suites of plant traits in variable environments. *Funct. Plant Biol.* **2009**, *36*, 665–681. [CrossRef] [PubMed]
22. Armstrong, J.; Armstrong, W. Rice: Sulphide-induced barriers to root radial oxygen loss, Fe^{2+} and water uptake, and lateral root emergence. *Ann. Bot.* **2005**, *96*, 625–638. [CrossRef]
23. Mittler, R. Oxidative stress, antioxidants and stress tolerance. *Trends Plant Sci.* **2002**, *7*, 405–410. [CrossRef]

24. Bhusal, N.; Kim, H.S.; Han, S.G.; Yoon, T.M. Photosynthetic traits and plant–water relations of two apple cultivars grown as bi-leader trees under long-term waterlogging conditions. *Environ. Exp. Bot.* **2020**, *176*, 104111. [CrossRef]
25. María, L.V.; Mignolli, F.; Aispuru, H.T.; Luis, A.M. Rapid formation of adventitious roots and partial ethylene sensitivity result in faster adaptation to flooding in the *aerial roots (aer)* mutant of tomato. *Sci. Hortic.* **2016**, *201*, 130–139. [CrossRef]
26. Loreti, E.; Van Veen, H.; Perata, P. Plant responses to flooding stress. *Curr. Opin. Plant Biol.* **2016**, 33, 64–71. [CrossRef] [PubMed]
27. Lei, S.T.; Zeng, B.; Xu, S.J.; Zhang, X.P. Response of basal metabolic rate to complete submergence of riparian species Salix variegata in the Three Gorges reservoir region. *Sci. Rep.* **2017**, *7*, 13885. [CrossRef]
28. Miao, L.F.; Xiao, F.J.; Xu, W.; Yang, F. Reconstruction of Wetland Zones: Physiological and biochemical responses of Salix variegata to winter submergence—A case study from water level fluctuation zone of the Three Gorges Reservoir. *Pol. J. Ecol.* **2016**, *64*, 45–52. [CrossRef]
29. Wang, C.Y.; Xie, Y.Z.; He, Y.Y.; Li, X.X.; Yang, W.H.; Li, C.X. Growth and physiological adaptation of *Salix matsudana koidz.* to periodic submergence in the hydro-fluctuation zone of the Three Gorges Dam Reservoir of China. *Forests* **2017**, *8*, 283. [CrossRef]
30. He, Y.Y.; Wang, C.Y.; Yuan, Z.X.; Li, X.X.; Yang, W.H.; Song, H.; Li, C.X. Photosynthetic characteristics of *Taxodium ascendens* and *Taxodium distichum* under different submergence in the hydro-fluctuation belt of the Three Gorges Reservoir. *Acta Ecol. Sin.* **2018**, *38*, 2722–2731. [CrossRef]
31. Wang, C.Y.; Li, C.X.; Wei, H.; Xie, Y.Z.; Han, W.J. Effects of long-term periodic submergence on photosynthesis and growth of *Taxodium distichum* and *Taxodium ascendens* saplings in the hydro-fluctuation zone of the Three Gorges Reservoir of China. *PLoS ONE* **2016**, *11*, e0162867. [CrossRef] [PubMed]
32. Li, C.X.; Wei, H.; Geng, Y.H.; Schneider, R. Effects of submergence on photosynthesis and growth of *Pterocarya stenoptera* (Chinese wingnut) seedlings in the recently-created Three Gorges Reservoir region of China. *Wetl. Ecol. Manag.* **2010**, *18*, 485–494. [CrossRef]
33. Magdziak, Z.; Mleczek, M.; Rutkowski, P.; Goliński, P. Diversity of low-molecular weight organic acids synthesized by Salix growing in soils characterized by different Cu, Pb and Zn concentrations. *Acta Physiol. Plant.* **2017**, *39*, 137. [CrossRef]
34. Adeleke, R.; Nwangburuka, C.; Oboirien, B. Origins, roles and fate of organic acids in soils: A review. *J. Bot.* **2017**, *108*, 393–406. [CrossRef]
35. Huang, G.Y.; Guo, G.G.; Yao, S.Y.; Zhang, N.; Hu, H.Q. Organic acids, amino acids compositions in the root exudates and Cu-accumulation in castor (*Ricinus communis* L.) Under Cu stress. *Int. J. Phytoremediat.* **2016**, *18*, 33–40. [CrossRef] [PubMed]
36. Yang, L.T.; Qi, Y.P.; Jiang, H.X.; Chen, L.S. Roles of organic acid anion secretion in aluminium tolerance of higher plants. *BioMed Res. Int.* **2013**, 173682. [CrossRef] [PubMed]
37. Wang, Y.; Chen, X.; Whalen, J.K.; Cao, Y.H.; Quan, Z.; Lu, C.Y.; Shi, Y. Kinetics of inorganic and organic phosphorus release influenced by low molecular weight organic acids in calcareous, neutral and acidic soils. *J. Plant Nutr. Soil Sci.* **2015**, *178*, 555–566. [CrossRef]
38. Wang, T.; Wei, H.; Zhou, C.; Chen, H.C.; Li, R. Responses of root organic acids and nonstructural carbohydrates of *Taxodium distichum* to water-level changes in the hydro-fluctuation belt of the Three Gorges Reservoir. *Acta Ecol. Sin.* **2018**, *38*, 3004–3013. [CrossRef]
39. Li, C.X.; Wei, H.; Lv, Q.; Zhang, Y. Effects of Water Stresses on Growth and Contents of Oxalate and Tartarate in the Roots of Chinese Wingnut (*Pterocarya stenoptera*) Seedlings. *For. Res.* **2010**, *46*, 81–88. [CrossRef]
40. Zhong, Y.; Liu, Z.X.; Qin, H.W.; Xiong, Y.; Xiang, L.X.; Liu, R.; Yang, Y.; Ma, R. Effects of winter submergence and waterlogging on growth and recovery growth of *Salix babylonica*. *J. South. Agric.* **2013**, *44*, 275–279.
41. Wang, T.; Wei, H.; Ma, W.C.; Zhou, C.; Chen, H.C.; Li, R.; Li, S. Response of *Taxodium distichum* to winter submergence in the water-level-fluctuating zone of the Three Gorges Reservoir region. *J. Freshw. Ecol.* **2019**, *34*, 1–17. [CrossRef]
42. Elcan, J.M.; Pezeshki, S.R. Effects of Flooding on Susceptibility of *Taxodium distichum* L. Seedlings to Drought. *Photosynthetica.* **2002**, *40*, 177–182. [CrossRef]
43. Arif, M.; Zhang, S.L.; Zheng, J.; Wokadala, C.; Mzondi, P.S.; Li, C.X. Evaluating the Effects of Pressure Indicators on Riparian Zone Health Conditions in the Three Gorges Dam Reservoir, China. *Forests* **2020**, *11*, 214. [CrossRef]
44. Ren, Q.S.; Song, H.; Yuan, Z.X.; Ni, X.L.; Li, C.X. Changes in soil enzyme activities and microbial biomass after revegetation in the Three Gorges Reservoir, China. *Forests* **2018**, *9*, 249. [CrossRef]
45. Yamashita, N.; Tanabata, S.; Ohtake, N.; Sueyoshi, K.; Sato, T.; Higuchi, K.; Ohyama, T. Effects of Different Chemical Forms of Nitrogen on the Quick and Reversible Inhibition of Soybean Nodule Growth and Nitrogen Fixation Activity. *Front. Plant Sci.* **2019**, *10*, 131. [CrossRef]
46. Park, S.; Cho, E.; Chung, H.; Cho, K.; Sa, S.; Balasubramanian, B.; Jeong, Y. Digestibility of phosphorous in cereals and co-products for animal feed. *Saudi J. Biol. Sci.* **2019**, *26*, 373–377. [CrossRef] [PubMed]
47. Gao, Z.X.; Zhou, G.M.; Huang, C.; Li, C.X.; Wang, L. Rapid determination of rudimental organic acids in root of Taxodium ascendens and *Taxodium distichum* by ion-suppression RP-HPLC. *Chin. J. Pharm. Anal.* **2005**, *25*, 1082–1085.
48. Huang, T.Z.; Wang, S.J.; Liu, X.M.; Liu, H.; Wu, Y.Y.; Luo, X.Q. Rapid determination of eight organic acids in plant tissue by sequential extraction and high performance liquid chromatography. *Chin. J. Chrom.* **2014**, *32*, 1356–1361. [CrossRef]
49. Field, A.; Miles, J.; Field, Z. *Discovering Statistics Using R*; SAGE Publications Ltd.: London, UK, 2012.

50. de Moraes Pontes, J.G.; Vendramini, P.H.; Fernandes, L.S.; de Souza, F.H.; Pilau, E.J.; Eberlin, M.N.; Magnani, R.F.; Wulff, N.A.; Fill, T.P. Mass spectrometry imaging as a potential technique for diagnostic of Huanglongbing disease using fast and simple sample preparation. *Sci. Rep.* **2020**, *10*, 13457. [CrossRef] [PubMed]
51. Yang, C.X.; Zhao, W.N.; Wang, Y.N.; Zhang, L.; Huang, S.C.; Lin, J.X. Metabolomics analysis reveals the alkali tolerance mechanism in puccinellia tenuiflora plants inoculated with arbuscular mycorrhizal fungi. *Microorganisms* **2020**, *8*, 327. [CrossRef]
52. Li, C.X.; Wei, H.; Lv, Q.; Zhang, Y. Effects of different water treatments on growth and contents of secondary metabolites in roots of slash pine (*Pinus elliottii Engelm.*) seedlings. *Acta Ecol. Sin.* **2010**, *30*, 6154–6162.
53. Liu, Y.H.; Peng, X.X.; Yu, L. Difference in oxalate content between buckwheat and soybean leaves and its possible cause. *J. Plant Physiol. Mol. Biol.* **2004**, *20*, 201–208. [CrossRef]
54. Zhang, J.; Gao, N.F.; Yang, H. Study on the organic acids during the grapes growing. *Liquor Mak.* **2004**, *34*, 69–71.

Article

Gap Size in Hyrcanian Forest Affects the Lignin and N Concentrations of the Oriental Beech (*Fagus orientalis* Lipsky) Fine Roots but Does Not Change Their Morphological Traits in the Medium Term

Alireza Amoli Kondori [1], Kambiz Abrari Vajari [1,*], Mohammad Feizian [1], Antonio Montagnoli [2] and Antonino Di Iorio [2]

[1] Faculty of Agriculture and Natural Resource, Lorestan University, Khorramabad 6815144316, Iran; amolikondori@gmail.com (A.A.K.); feizian.m@lu.ac.ir (M.F.)
[2] Department of Biotechnology and Life Science, University of Insubria, 21100 Varese, Italy; antonio.montagnoli@uninsubria.it (A.M.); antonino.diiorio@uninsubria.it (A.D.I.)
* Correspondence: kambiz.abrari2003@yahoo.com; Tel.: +98-911-344-1571

Abstract: *Research Highlights:* Fine roots play an important role in plant growth as well as in carbon (C) and nutrient cycling in terrestrial ecosystems. Gaining a wider knowledge of their dynamics under forest gap opening would improve our understanding of soil carbon input and below-ground carbon stock accumulation. Single-tree selection is increasingly recognized as an alternative regime of selection cutting sustaining biodiversity and carbon stock, along with timber production, among ecosystem functions. However, the fine root response in terms of morphological and chemical composition to the resulting harvest-created gaps remains unclear. *Background and Objectives:* This paper investigates the effect in the medium term (i.e., 6 years after logging) of differently sized harvest-created gaps on fine root dynamics and chemical composition. *Materials and Methods:* A total of 15 differently sized gaps (86.05–350.7 m^2) and the adjacent 20 m distant closed canopies (control) were selected in a temperate *Fagus orientalis* forest (Hyrcanian region, Iran). Eight soil cores were collected at the cardinal points of the gap edge, including four facing the gap area—the same at the adjacent intact forest. *Results:* For the selected edge trees, the different size of gaps, the core position, and the tree orientation did not affect the investigated morphological traits, except for the slightly higher specific root length (SRL) for the larger fine root fraction (1–2 mm) in the side facing the gap area. Differently, the investigated chemical traits such as N concentration and cellulose:lignin ratio significantly increased with increasing gap size, the opposite for C:N ratio and lignin. Moreover, N concentration and C:N significantly decreased and increased with the fine root diameter, respectively. *Conclusions:* This work highlighted that, in the medium term and within the adopted size range, artificial gap opening derived from single-tree selection practice affected the chemistry rather than the biomass and morphology of gap-facing fine roots of edge trees. The medium term of six years after gap creation might have been long enough for the recovery of the fine root standing biomass to the pre-harvest condition, particularly near the stem of edge trees. A clear size threshold did not come out; nevertheless, 300 m^2 may be considered a possible cut-off determining a marked change in the responses of fine roots.

Keywords: forest gap; forest management; fine roots; morphology; lignin; carbon; nitrogen; *Fagus orientalis*

Citation: Kondori, A.A.; Vajari, K.A.; Feizian, M.; Montagnoli, A.; Di Iorio, A. Gap Size in Hyrcanian Forest Affects the Lignin and N Concentrations of the Oriental Beech (*Fagus orientalis* Lipsky) Fine Roots but Does Not Change Their Morphological Traits in the Medium Term. *Forests* **2021**, *12*, 137. https://doi.org/10.3390/f12020137

Academic Editor: Antonio Montagnoli
Received: 16 December 2020
Accepted: 21 January 2021
Published: 26 January 2021

1. Introduction

Fine roots (diameter <2 mm) are the most dynamic and sensitive component within the overall root system [1,2], playing a crucial role in water and nutrient acquisition [3]. They have also been regarded as short-lived and recognized as the most important component contributing to below-ground C and N fluxes in forest ecosystems [4,5]. Indeed, estimation of fine root production accounts for as much as 33–67% of the total annual net primary production (NPP) [6], although fine root biomass contributes relatively little to total forest biomass, usually <5% [7]. Furthermore, fine root mortality contributes from 18 to 58% of total N to forest soils, making the root litter N addition higher than that from above-ground litterfall in some ecosystems [4,5].

Fine root biomass has been found to vary with forest stand characteristics, i.e., tree species, stand age, density, basal area, canopy openness and soil properties, or environmental factors such as air temperature, amount of precipitation, geographical location, and elevation. In general, changes in canopy openness resulting from differently shaped gaps or coppice conversion to high forest highlighted a significant decrease in fine root biomass in the exposed soil [8–10]. For both types of management practices, fine root biomass has been found to decrease in the exposed soil surface of the gap center and proportionally to its increasing size [9,11], and reduced basal area [12]. Morphological traits related to fine root biomass also vary with gap opening since the mean diameter of the fine root population changes with change in the soil moisture caused by the opening in the canopy cover [13,14]. For example, under drier soil conditions, plants produce longer and finer roots [13,15], which results in a relatively greater length per unit mass, thereby leading to an increase in specific root length (SRL, length-to-mass ratio). Concerning the chemical traits related to the fine root biomass, carbon and nitrogen concentrations are well known to change with changes in diameter classes, branching order [16,17], and seasonality [16,18]. In beech forests, for example, fine root C and N concentrations show a reverse pattern compared to each other, with the highest and lowest values for C concentration in July and October, respectively [18]. However, how fine root morphological and chemical traits vary with gap size is still poorly investigated.

Forest ecosystems are exposed to natural disturbances, which can have a serious impact on their functioning especially in a worsening global climate scenario [19]. In particular, disturbances such as diseases [20–22], storm [23,24], fire [25] may cause injuries to trees enough to cause the formation of forest gaps. In temperate broadleaf forests, where large-scale disturbances are rare, natural regeneration occurs predominantly in gaps [26,27]. Therefore, forest management that approximates nature, such as single-tree selection practice, appears to be a flexible and useful tool to secure sustainable forest development in terms of biodiversity [26,27] and uneven-aged forests [28]. This tool mimics natural openings of various sizes that follow moderate disturbance events [29]. On the other hand, both natural and artificial gaps induce the alteration of microclimate conditions (i.e., soil moisture and temperature, irradiance) either inside or in the proximity of the gaps [30], leading to below-ground responsive adjustments in terms of fine root dynamics whose magnitude may change with the increasing size of the gap [11]. Therefore, unveiling the threshold that could lower the impact on fine root morphological and chemical traits of edge trees may be of interest for sustainable forest management, which would lead to a higher amount and lasting in time below-ground carbon stock accumulation. This is particularly relevant as no common agreement exists on the definition of the size class of the gap; for example, the small criterion ranges from 300 [29], 700 [31,32] to 1250–1960 m^2 [29].

Differences in fine root biomass among forest stands have been modeled to change also from stand initiation to a later stage of stand development (canopy closure) [33]. In particular, short-lived fine roots generally <0.5 mm in diameter [34,35] and characterized by higher turnover rate are predominant during the first years following thinning operations [36,37]. For this reason, short-term (1–4 years) studies may not adequately track the fine root fraction characterized by a higher longevity and that mostly contribute to the standing biomass [36,37]. Thus, choosing the sampling time since the gap opening occurred is crucial for a correct estimation of the fine root standing biomass, and medium-term investigation may be considered a good compromise.

Fine root form and function may differ among branching orders [38,39] and the shortcomings of the diameter-based approach are widely recognized [40]. Therefore, in this study, the size class approach was maintained for a better discrimination between the

<0.5 mm diameter class (i.e., very fine roots), which represent the most dynamic component of the root system, and the 0.5–1 and 1–2 mm diameter classes, which represent the more stable and woodier portion [41–43].

Oriental (or eastern) beech (*Fagus orientalis* Lipsky) forests represent the most important tree cover in the temperate-broadleaved forest in northern Iran, playing a key role in forestry activities. Caspian forests of Iran have been harvested under different silvicultural systems, such as shelterwood cutting or single-tree selection, which leads to different canopy openness [19]. Single-tree selection is increasingly recognized as an alternative regime of selection cutting sustaining biodiversity, recruitment and carbon stock, along with timber production, among ecosystem functions [44]. Many studies were conducted about different features of *F. orientalis* trees, but few on the fine roots [31,32] and their response at the morphological and chemical level to the different-sized gaps.

In this study, the two-fold hypothesis was that, in the medium term, gap openings might: (i) induce a reduction in fine root biomass greater than in length and, consequently, increase the specific root length (SRL), and (ii) increase the metabolism of fine roots leading to a higher N concentration. These variations would be of greater magnitude with the increasing gap size. To test these hypotheses, for trees positioned at the edge of different-sized gaps, fine root biomass, length, and tissue density were measured together with C, N, lignin, and cellulose concentrations. This approach was intended to use the *F. orientalis* tree fine root morpho-chemical traits as indicators of a threshold gap size above which a significant influence on tree ecophysiology may occur.

2. Materials and Methods

2.1. Study Area

To study the effects of gap size on fine roots properties, a broadleaved *Fagus orientalis* L. forest, covering an area of 40.4 ha, was selected in the Caspian area (Mazandaran province), northern Iran (36°12′ N,53°24′ E). The general characteristics of this area were already described in a previous work carried out on the same site [45]. Briefly, the study site is located at an elevation of 1000–1200 m a.s.l., on a north-facing slope of 0–30%. The climate is humid with a mean annual temperature of 10.5 °C and a mean annual precipitation of 858 mm. The pseudogleyic and gley are the dominant soil types. The shade-adapted and broadleaved *F. orientalis* trees occupy all canopy layers, from overtopped to dominant ones, and other tree species including hornbeam (*Carpinus betulus*), alder (*Alnus subcordata*), and maple (*Acer velutinum*) are also present. The oriental beech stand in the study site is multi-layered, uneven-aged, and developed from natural regeneration. At the time of fine root sampling, the stand density was 178 trees ha^{-1}, with a mean tree height and diameter at breast height (DBH) of 30.95 ± 0.79 m (mean ± SE) and 58.83 ± 2.37 cm, respectively. No natural disturbance regime, such as storm, heavy snow or fire, is recorded in the forestry plan. The silvicultural practice historically adopted is the single-tree selection, which produced different-sized artificial gaps depending on the crown size of the harvested tree.

2.2. Experimental Design

In 2011, 45 gaps were randomly opened on the stand by felling different-sized trees. A 50 m buffer of trees was maintained between all gaps, equal to twice the diameter of the largest gap. Forest harvesting operations were conducted to prevent soil disturbance as much as possible within the stand. In total, 15 elliptical different-sized gaps ranging from 86.05 to 350.7 m^2 were selected for this study. Detailed information on gap and related tree sizes are reported in Table S1. Each single gap was considered as the experimental unit, making this experimental design a point comparison approach rather than a replicated experiment on the ecosystem scale. The gap area was measured using the formula for an ellipse: $A = (\pi LW)/4$, where L and W (m) are the longest and the largest perpendicular to L distances within the gap, respectively (Figure 1). These distances were measured between stems of the edge trees. All gaps were oriented with the short distance W always parallel to the north direction. To compare with undisturbed forest, the adjacent 20 m distant closed-canopy clustered trees (uncut control) were selected for each gap size.

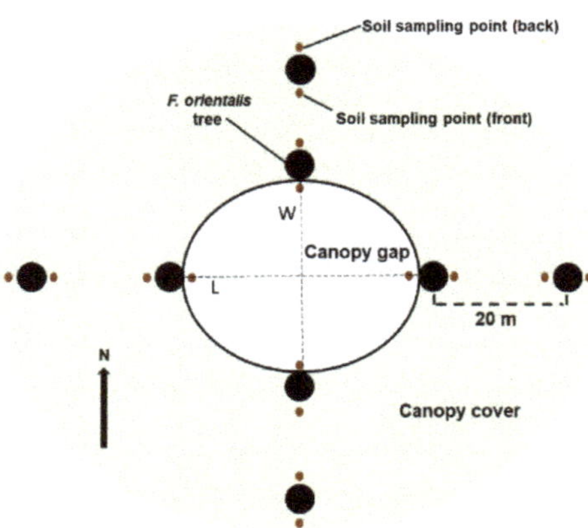

Figure 1. Sketch of location of the sampling points in the gap (white area) and adjacent intact forest (light green area). Large black circles indicate the sampled trees, small brown circles the soil core position, in the front and back of the selected trees. Dashed L and W lines indicate the distances among stems used for ellipse calculation. All gaps were oriented with the short distance of the ellipse always parallel to the north direction.

2.3. Fine Root Sampling

Field sampling of fine roots was carried out late in the growing season in October 2017 in all 15 artificial gaps and adjacent intact forest, i.e., 6 years after gap creation. Soil samples were collected by hand soil corer (Root Auger 80 mm inner diameter, ELE International, Bedfordshire, UK) to 20 cm soil depth at 1 m distance from the trunk. The sampling protocol followed the scheme reported in Figure 1 [41]. For each gap, eight soil cores were collected at the cardinal points of the gap border, and eight at the adjacent intact forest at 20 m from the gap. Of the eight cores, four were sampled in the side facing the gap area (front) and four on the opposite side (back). Core samples were stored in plastic bags in a commercial cool box (mod. 9315, Gio Style Spa, Italy) including ordinary freezer packs, transported to the laboratory, and kept at 4 °C until processed.

2.4. Morphological Features of Fine Roots

In the laboratory, all soil samples were washed on a sieve (1 mm mesh size) to remove the soil. Soil freed fine roots were further cleaned from soil residues under a stereo-microscope, and beech fine roots were distinguished from other understory roots by morphological characteristics. Beech fine roots appear reddish and stiffer than the understorey roots; these morphological characteristics were previously established from samples dug near the tree. Fine root samples were then scanned at a resolution of 500 dpi with a calibrated flatbed scanner coupled to a lighting system (Expression 10,000 XL, Epson America Inc., Long Beach, CA, USA). The resulting images were analyzed with WinRhizo Pro V. 2007d software (Regent Instruments Inc., Quebec, QC, Canada), which, setting the diameter classes with different colors, made it possible to group with high accuracy the root axes in three diameter classes (<0.5; 0.5–1.0; 1.0–2.0 mm); root axes were separated from each other where necessary with scissors or scalpels. The morphometric measurements as length and mean diameter were performed. Successively, fine-roots belonging to each diameter class were grouped, oven-dried at 70 °C (48 h), and weighed for dry mass measurements. Morphometric data, together with dry weight data, were also used to calculate the relative morphological traits specific root length (SRL, m g^{-1}) and root tissue density (RTD, g cm^{-3}).

2.5. Chemical Composition of Fine Roots

In our samples, as the dry weight per core sample for the three fine root diameter classes ranged from 0.05 to 0.87 g, the amount was not sufficient for testing the chemical properties, particularly for very fine roots. Then, in the laboratory, for each diameter class, all eight cores for each gap size were pooled together, the same for the closed canopy, resulting in one sample per diameter class per gap size, i.e., 15 per diameter class in total. Successively, only 9 gap sizes were selected by eliminating the even values from the series of 15 gaps (2, 4, 6 ... sample). In detail, every even sample was equally split, one half pooled with the previous sample and the second half with the successive one. In this way, the critical amount of 2–3 g per sample has been achieved, totaling 18 samples corresponding to 9 gaps and 9 closed canopies for each diameter class (54 in total). Each of the 54 samples was ground in liquid nitrogen with mortar and pestle and milled to pass a 40-mesh (37 μm mesh opening) screen; the powder was used for carbon, nitrogen, lignin, cellulose, and phosphorus concentrations measurement.

2.5.1. Carbon and Nitrogen

C and N concentrations were measured with a CHN analyzer (NA-2000 N-Protein; Fisons Instruments S.p.A., Rodano [MI], Italy). The analyzer was calibrated with an atropine standard (Sigma-Aldrich, A-0132, St. Louis, Missouri, MO, USA) and every 10th sample with an atropine sample. The mean total N and C recovery rate for nutrient analysis of atropine was 100.28 ± 0.59% and 100.62 ± 0.22%, respectively.

2.5.2. Cellulose and Lignin

The method used to measure total cellulose content was based on that developed by Leavitt and Danzer [46,47] and consisted of removing as many non-cellulosic compounds as possible from the root material. The first compounds to be removed were lipids (waxes, oils, and resins). Each sample was poured into a Teflon sachet (no. 11803, pore size 1.2 μm) and in groups of nine placed into a soxhlet extractor (50 mm i.d., 200 mL capacity to siphon top, mod. 64826, Supelco | Sigma-Aldrich) equipped with a flask containing a 700 mL mixture of toluene 99%–ethanol 96% (2–1; v/v) heated until boiling point. After 24 h of extraction, the extractor solution was replaced with 700 mL of ethanol heated to the same temperature. After 24 h, the samples were removed from the soxhlet and immersed in distilled water heated to 100 °C for 6 h. This process removes hydrosoluble molecules from the sample. The second compounds to be removed were lignin. All samples were placed in a 250 mL beaker containing 160 mL of distilled water, 1.5 g of sodium chlorite ($NaClO_2$), and 0.5 mL of acetic acid. The sample and solution were shaken using a magnetic stirrer and heated to 70–80 °C for 1 h (this procedure was repeated three times, 3 h in total). After the flask was cooled to a constant temperature, the sample was removed and filtered using a filter flask and washed with distilled water until it was free from acid. The samples were dried at ambient temperature during 12 h and weighed. The percentage of cellulose was evaluated by calculating the relative difference in the initial and final weight of each sample (0.5 g).

Lignin content was measured according to a previous protocol [48], with minor modifications. For lignin extraction, 1 g of powdered sample was poured in a 15 mL plastic falcon and boiled with 2 mL ethanol for 30 min and left overnight on a tilting plate. After centrifugation, Falcons were centrifuged at 10,000× g for 15 min at 25 °C, the supernatant removed and the pellet homogenized in 10 mL of extraction buffer (50 mM Tris–HCl, 0.01% Triton X-100 (10 g L^{-1}), 1 M NaCl, pH 8.3). The suspension was stirred, centrifuged at 10,000× g for 10 min, washed twice with 4 mL of extraction buffer, twice with 2 mL of 80% (v/v) acetone, and twice with 2 mL of acetone, and then dried in a concentrator. Each pellet was then treated with 0.4 mL of thioglycolic acid and 2 mL of 2 M HCl, for 4 h, at 95 °C, centrifuged at 15,000× g for 10 min and washed three times with distilled water. Lignothioglycolic acid from each pellet was then extracted with 2 mL of 0.5 M NaOH, under stirring for 16 h, at 25 °C. The supernatant was acidified with 0.4 mL of 37% (v/v) HCl in proportion 1:5 to the sample volume. Lignothioglycolic acid was then precipitated at 4 °C, for 4 h, recovered by centrifugation at 15,000× g for 20 min, and dissolved in 1 mL of 0.5 M NaOH. The lignin amount within each sample was calculated by measuring the absorbance at 280 nm, using a specific absorbance coefficient of 6.0 L g^{-1} cm^{-1}. Because this specific absorbance coefficient provides only an approximate conversion (the absorbance of lignothioglycolic acid from different sources can vary considerably; see Doster and Bostock [48]), all readings were normalized to the specimen with the highest lignin content [49].

2.6. Statistical Analyses

To compare the different gap sizes and the difference between the fine roots facing the gap and those facing the opposite side for each cardinal point, 15 artificial gaps of increasing size and 15 adjacent 20 m distant uncut control plots were established. In this work, no a priori subjective thresholds, such as small, medium and large sizes, have been adopted within the considered gap size range (86.05–350.70 m^2), so gap size was used as a covariate in the analysis of covariance (ANCOVA). Two-way ANCOVA was performed to evaluate the effects of the fixed factor core position (front, back) and edge tree orientation (N, S, E, W), as well as their interaction, and the covariate gap size on several morphological traits of fine roots with diameters of <0.5, 0.5–1.0 and 1.0–2.0 mm. For chemical traits (Section 2.4), as fine root samples were pooled, a one-way ANCOVA was performed with the diameter classes as a fixed factor and the gap size as a covariate. All uncut control data were pooled as one and treated as a control.

Data were square-root or log-transformed where necessary to meet normality and homoscedasticity assumptions. Data given in figures are not transformed; p and R^2 values of regression analysis refer to data transformed where necessary. All statistical analyses were performed with SPSS version 20.0 (IBM, Armonk, NY, USA) software and were performed with a 5% rejection level.

3. Results and Discussion

3.1. Morphological Traits

For the edge trees, the different size of gaps did not affect the investigated morphological traits, neither the core position nor the tree orientation (Table 1). If pooled, the mean values of all morphological traits did not differ from those of the uncut control (closed canopy), independently from the size of gaps (Figure 2); some deviations either above or below the control mean occurred in the center-left of the distribution, i.e., for the smaller gap sizes. The only exception was the slightly higher SRL for the larger fine root fraction (1–2 mm) in the side facing the gap area (Figure 3), which marginally missed the significance (core position $p = 0.071$; Table 1).

Table 1. F and p values of ANCOVA (General Linear Model) for the effects of core position and orientation on morphological fine root traits divided by diameter classes. Gap sizes were used as covariates. Interactions were not significant and therefore excluded from the model.

Fine Root Trait	Diameter (mm)	Core Position (df = 1)		Orientation (df = 3)		Gap Size (c) (df = 1)	
		F	p	F	p	F	p
Length (m m^{-2})	0.5	0.125	0.724	0.343	0.794	0.087	0.768
	0.5–1	0.054	0.817	0.031	0.993	0.152	0.697
	1–2	0.274	0.601	0.163	0.921	0.510	0.476
Dry mass (g m^{-2})	0.5	0.009	0.924	0.329	0.805	0.007	0.935
	0.5–1	0.067	0.796	0.338	0.798	0.030	0.862
	1–2	0.843	0.360	0.207	0.891	0.253	0.616
RTD (g cm^{-3})	0.5	0.405	0.525	0.387	0.762	0.137	0.712
	0.5–1	0.096	0.756	1.220	0.303	1.387	0.240
	1–2	1.082	0.299	0.822	0.483	0.844	0.359
SRL (m g^{-1})	0.5	0.687	0.408	0.434	0.729	0.804	0.371
	0.5–1	0.001	0.998	0.671	0.571	0.659	0.418
	1–2	3.296	0.071	1.230	0.299	0.135	0.714

RTD, Root Tissue Density; SRL, Specific Root Length; (c), covariate

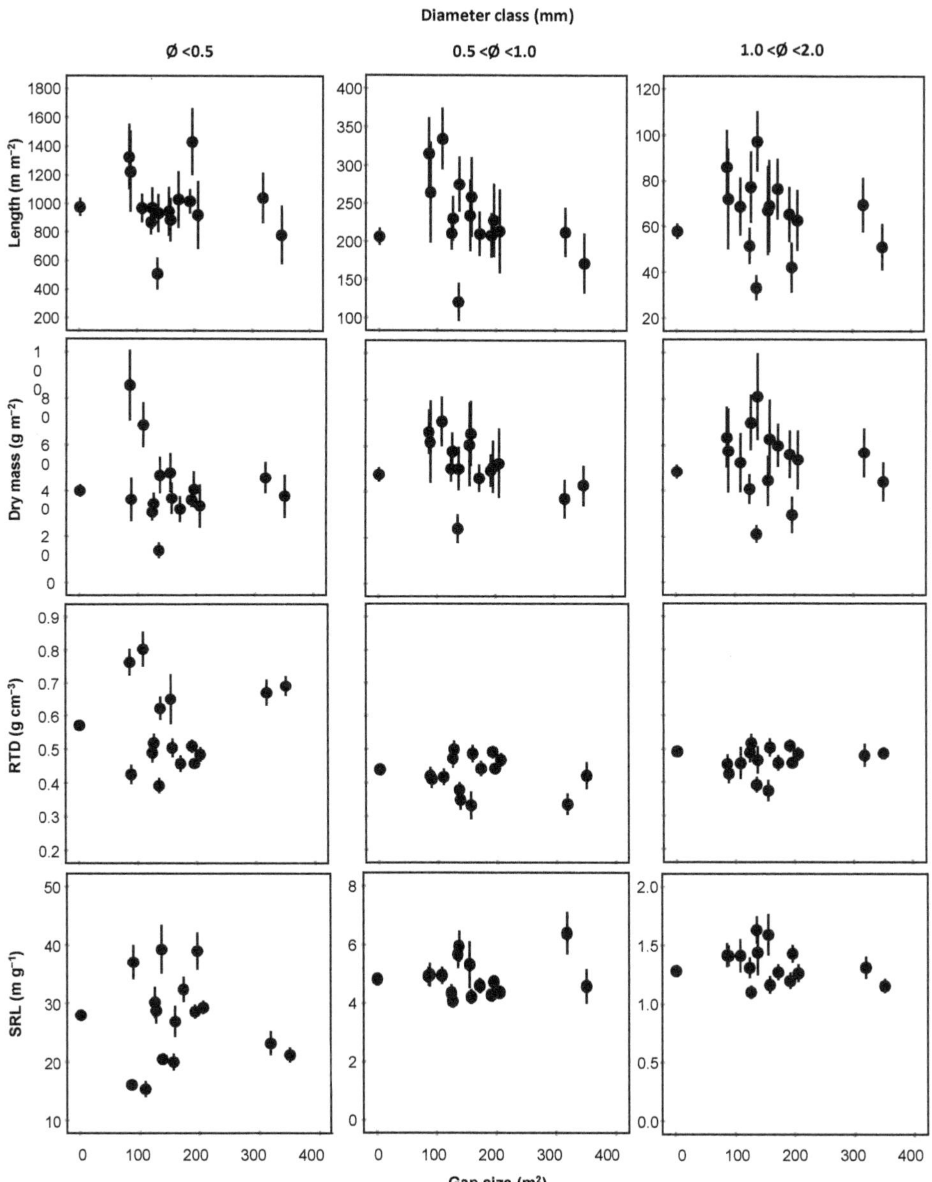

Figure 2. Fine root length, dry mass, root tissue density (RTD), and specific root length (SRL) (columns), according to three diameter classes (<0.5, 0.5–1, and 1–2 mm) (rows), in relation to gap size. Each point represents the mean of 8 replicates (front and back position and orientation pooled together) ± SE. The uncut control was conventionally assigned the gap size value of 1 m^2 and was the mean of 120 replicates (front and back position, orientation, and 15 gap size pooled together). If not reported, the scale for a given variable is the same for all the three panels.

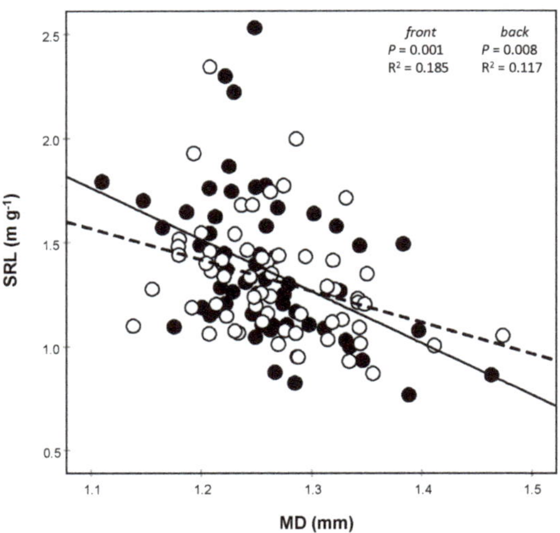

Figure 3. Relationship between specific root length (SRL) and mean diameter (MD) of the 1–2 mm diameter fine root class according to the front (filled circle) and back (empty circle) core positions. Each point represents the mean of 4 replicates ± SE. Continuous and dashed lines represent front and back linear interpolations, respectively.

The mean diameter (MD) (Figure 3) rather than the RTD (data not shown) contributed to this slightly higher gap facing, 1–2 mm in diameter SRL, as the relationship was stronger for the gap-facing side. Moreover, the average diameter slightly increased with increasing gap size (Figure S1). Therefore, this result does not support the first hypothesis except for the slightly higher SRL of the larger fine root fraction (1–2 mm).

Unfortunately, no data in the short-term have been collected, making any consideration on fine root dynamics over time merely speculative. Nevertheless, the decrease in fine root biomass following gap openings is supported by other short-term experiments [8,9] (6 and 11 months after logging, respectively), which found a consistent decrease at the gap edge, and almost no growth in the center of the gap compared to the adjacent intact forest. Furthermore, regarding the spatial localization of the soil sampling points for edge trees, the adopted distances of 5 and 8 m from the trunk in the same short-term experiments [8,9] were higher than the 1 m used in the present work, which fell under the canopy crowns of rather tall trees (on average 31 m). In fact, previous findings on the soil characteristics (moisture content, bulk density, total N, P, and soil organic carbon) from the same experimental gaps [45] showed a lack of significant differences between edge and closed canopy trees. These findings also highlighted the low impacts of the adopted management practices on soil characteristics, which did not extend over the medium term [50,51]. This scanty soil alteration might explain the lack of significance of the orientation factor for the edge trees, although a higher soil temperature should be expected for south-faced trees. Therefore, the fine root biomass of edge trees might have been scarcely reduced after gap openings, and the medium-term observation adopted in this work may have been long enough to enable the recovery of fine root biomass to pre-harvest levels, explaining the lack of morphological differences among the gap sizes investigated.

3.2. Chemical Traits

Interestingly, N concentration significantly ($p < 0.001$) increased with increasing size of gaps and significantly ($p < 0.001$) decreased with increasing fine root diameter (Table 2; Figure 4). C concentration did not show any trend in response to the different sizes of the gap, whereas it marginally decreased ($p = 0.096$) with increasing root diameter (Table 2; Figure 4). C:N ratio marginally decreased ($p = 0.087$) with increasing gap size, whereas it significantly ($p < 0.001$) increased with increasing fine root diameter (Table 2, Figure 4). It is noteworthy that having pooled the samples did not withhold the clear trend obtained for N concentration.

Table 2. F and p values of ANCOVA (General Linear Model) for the effects of diameter classes on chemical fine root traits. Gap size was used as a covariate. Interactions were not significant and therefore excluded from the model. Boldface p values are significant at a probability level of $p < 0.05$.

Chemical Trait	Diam Class (df = 2)		Gap Size (c) (df = 1)	
	F	p	F	p
C	2.598	0.096	1.596	0.219
N	43.73	<0.001	25.28	<0.001
C:N	20.89	<0.001	3.05	0.087
Cellulose	0.374	0.692	0.024	0.879
Lignin	1.747	0.197	9.371	0.006
Cellulose:Lignin	2.299	0.123	13.442	0.001
Lignin: N	3.209	0.059	14.54	0.001

(c), covariate.

These findings are in accordance with the literature which reports a strong inverse correlation between N concentration and root diameter with the highest concentrations in the thinnest root portions [17,34], whereas no consistency emerges on the relationship between C concentration and root diameter [34,52] and references therein. Differently, a lack of consistency persists about the possible N concentration increase in fine roots in response to gap opening [30], particularly when consequent to artificial gap formation. Most of the studies had focused on alteration on soil processes such as nutrient release during litter decomposition [53,54], microbial activity [45,54], net mineralization and nitrification [55], but few papers concern fine roots [8,9,11]. Thinning operations stimulate the N concentration increase in European beech forests in the Southern Alps, [18], which results in fine roots with a shorter lifespan than those living in the forest left to grow for many years. Findings from the present study suggest a similar response at the fine root level to artificial gap opening derived from single-tree selection practice. Indeed, the lower C:N ratio observed in larger gaps independently of the diameter class highlighted the lower construction costs and, consequently, the more ephemeral nature of these fine roots [34]. Although not differentiated between orientation and core position, these chemical trends would reveal the stimulation of the fine root growth with the increasing size of the gap, with traces still present six years after gap opening.

For cell wall chemical compounds, lignin concentration did not change between diameter classes, but significantly decreased with increasing gap size only for the larger sub-classes, 0.5–1 and 1–2 mm (Table 2, Figure 4). Cellulose also did not differ between the diameter classes and decreased with increasing gap size, but only for the 1–2 mm diameter class. Interestingly, the Cellulose to lignin ratio resulted significantly higher in larger gaps only for the 0.5–1 and 1–2 mm classes, whereas lignin to N ratio decreased significantly with increasing gap size for the smaller <0.5 and larger 1–2 mm classes, and slightly increased with increasing diameter ($p = 0.059$).

The decrease in the cellulose and lignin in the larger fraction of fine roots with increasing gap size is indicative of roots with thinner secondary walls, and the finding of the slightly higher SRL and lower mean diameter in gap facing 1–2 mm fine roots (Figure 3) fits to this explanation. Moreover, the increasing trend of the cellulose:lignin ratio with increasing gap size is indicative of a lignin reduction proportionally higher than that of cellulose; the latter, in particular, did not significantly change with the different gap sizes if the last size value was removed from calculation. Xylem percentage area [30] and cellulose:lignin ratio [56–58] increase with the secondary growth and, consequently, increasing diameter. Conversely, RTD decreases with increasing root diameter [59,60] or xylem percentage area [30], although other studies have provided inconsistent results on these relationships [61]. Thus, the slight increase in the average diameter for the 1–2 mm class with the increasing size of the gap (Figure S1) marginally correlated with the increasing trend of the cellulose:lignin ratio. Readjustments in crown closure are well known to increase the radial growth in stem and structural roots through the enhancement of photosynthate production [12,62,63], and may have contributed to the moderate 1–2 mm root class radial growth.

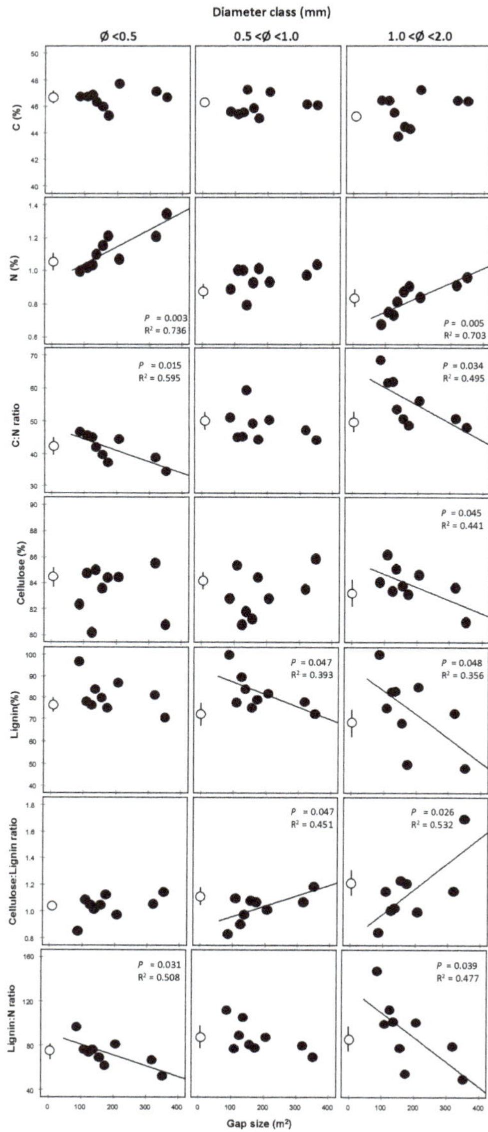

Figure 4. Fine root chemical traits (rows) in relation to gap size, according to three diameter classes (<0.5, 0.5–1, and 1–2 mm) (columns). Points represent one value per 9 gap sizes only (see Section 2.4); the uncut control, not included in the regression analysis, was conventionally assigned the gap size value of 1 m^2 and represents the mean of 9 replicates ± SE (see Section 2.5). If significant at $p < 0.05$, regression lines and the corresponding R^2 were shown.

The cellulose:lignin and Lignin:N ratios, in particular, also belong to those chemical traits of plant litter that have the highest impact on the decomposition rates. The lower Lignin:N ratio with increasing gap size found in the present work could result in higher decomposition rates [64–66]. However, simple extrapolation from the decomposability of root litter to, for example, the long-term carbon sequestration in forest soils is not possible, as additional factors such as the spatial inaccessibility of soil organic matter and organo-mineral interactions cannot be ruled out (e.g., von Lutzow et al. [67]). Thus, differently from the center of gaps, it may be assumed that stimulation of growth near the edge trees was not as remarkable at the morphological level as at it was at the chemical level. This apparent discrepancy between the morphological and chemical traits is explainable in that most of the studies on the effects of gap size on tree fine root dynamics refer to the short-term period, whereas in the present study, the medium term might have veiled the morphological effects not as much as the nitrogen and lignin content in the larger fraction. Morphological and chemical traits are frequently decoupled at the root system level [68]. Indeed, phenotypic plasticity was found to be limited across soil conditions and growing seasons for several temperate species [16,69,70]. Similarly, a correlation between morphological traits such as SRL and N concentration was lacking for fine roots of many softwood and hardwood North American species [68,71].

4. Conclusions

This work highlights that, in the medium term and within the adopted size range, artificial gap opening derived from single-tree selection practice affected the chemistry rather than the biomass and morphology of gap-facing fine roots in oriental beech edge trees. These outcomes suggest that the below-ground carbon stock is not influenced in the medium term by the forest gap openness following single-tree selection practice, but readjustments in the crown closure of edge trees may contribute to a moderate radial growth in the larger and woodier fine root fraction. Consequently, the derived increase in C:N and decrease in Lignin:N ratios with increasing gap size may increase the fine root decomposition, and subsequently the carbon input into the soil as medium-long-term implication. A clear size threshold did not come out since the trends with increasing gap size were either absent for the morphological or continuous for the chemical traits. Nevertheless, for this latter, 300 m^2 may be considered a possible cut-off determining a marked change in the responses of fine roots.

Supplementary Materials: The following are available online at https://www.mdpi.com/1999-4907/12/2/137/s1, Figure S1: Specific root length (SRL) and mean diameter (rows) in relation to gap size of the 1–2 mm diameter fine root class, Table S1: Size of the 15 studied gaps and the related tree-edge characteristics.

Author Contributions: Conceptualization, K.A.V., A.A.K., M.F., and A.D.I.; methodology, K.A.V., A.A.K., M.F., A.M., A.D.I.; investigation, A.A.K., M.F., K.A.V.; software, A.D.I., A.A.K. and A.M.; formal analysis, A.D.I., A.A.K. and A.M.; data curation, A.D.I., A.A.K.; writing—original draft preparation, A.A.K., K.A.V.; writing—review and editing, A.D.I., A.M.; visualization, A.D.I.; supervision, K.A.V., A.D.I.; funding acquisition, K.A.V., A.D.I. All authors have read and agreed to the published version of the manuscript.

Funding: This study was supported by Lorestan University, Lorestan, Iran and the University of Insubria (FAR no. 2019).

Data Availability Statement: The datasets generated during the current study are available from the corresponding author on reasonable request.

Acknowledgments: The authors are grateful to Lorestan University, Lorestan, Iran, and the University of Insubria (University Research Funding—project FAR) to provide us with financial supports. Special thanks to Enrico Caruso for his technical support with the Soxhlet system for cellulose quantification, Delle Fratte Michele for assistance with the CHN analysis. The authors gratefully acknowledge the two anonymous reviewers for their valuable comments.

Conflicts of Interest: The authors declare no conflict of interest.

References

1. Wang, C.; Brunner, I.; Zong, S.; Li, M.-H. The Dynamics of Living and Dead Fine Roots of Forest Biomes Across the Northern Hemisphere. *Forests* **2019**, *10*, 953. [CrossRef]
2. Finér, L.; Ohashi, M.; Noguchi, K.; Hirano, Y. Fine root production and turnover in forest ecosystems in relation to stand and environmental characteristics. *For. Ecol. Manag.* **2011**, *262*, 2008–2023. [CrossRef]
3. Železnik, P.; Vilhar, U.; Starr, M.; De Groot, M.; Kraigher, H. Fine root dynamics in Slovenian beech forests in relation to soil temperature and water availability. *Trees* **2016**, *30*, 375–384. [CrossRef]
4. Finér, L.; Ohashi, M.; Noguchi, K.; Hirano, Y. Factors causing variation in fine root biomass in forest ecosystems. *For. Ecol. Manag.* **2011**, *261*, 265–277. [CrossRef]
5. Yuan, Z.; Chen, H.Y. Fine root biomass, production, turnover rates, and nutrient contents in boreal forest ecosystems in relation to species, climate, fertility, and stand age: Literature review and meta-analyses. *Crit. Rev. Plant. Sci.* **2010**, *29*, 204–221. [CrossRef]
6. Jackson, R.B.; Mooney, H.A.; Schulze, E.D. A global budget for fine root biomass, surface area, and nutrient contents. *Proc. Natl. Acad. Sci. USA* **1997**, *94*, 7362–7366. [CrossRef]
7. Vogt, K.A.; Vogt, D.J.; Bloomfield, J. Analysis of some direct and indirect methods for estimating root biomass and production of forests at an ecosystem level. *Plant Soil* **1998**, *200*, 71–89. [CrossRef]
8. Bauhus, J.; Bartsch, N. Fine-root growth in beech (*Fagus sylvatica*) forest gaps. *Can. J. For. Res.* **1996**, *26*, 2153–2159. [CrossRef]
9. Brockway, D.G.; Outcalt, K.W. Gap-phase regeneration in longleaf pine wiregrass ecosystems. *For. Ecol. Manag.* **1998**, *106*, 125–139. [CrossRef]
10. Montagnoli, A.; Terzaghi, M.; Di Iorio, A.; Scippa, G.S.; Chiatante, D. Fine-root seasonal pattern, production and turnover rate of European beech (*Fagus sylvatica L.*) stands in Italy Prealps: Possible implications of coppice conversion to high forest. *Plant Biosyst.* **2012**, *146*, 1012–1022. [CrossRef]
11. Jones, R.H.; Mitchell, R.J.; Stevens, G.N.; Pecot, S.D. Controls of fine root dynamics across a gradient of gap sizes in a pine woodland. *Oecologia* **2003**, *134*, 132–143. [CrossRef] [PubMed]
12. Di Iorio, A.; Montagnoli, A.; Terzaghi, M.; Scippa, G.S.; Chiatante, D. Effect of tree density on root distribution in Fagus sylvatica stands: A semi-automatic digitising device approach to trench wall method. *Trees Struct. Funct.* **2013**, *27*, 1503–1513. [CrossRef]
13. Ostonen, I.; Püttsepp, Ü.; Biel, C.; Alberton, O.; Bakker, M.R.; Lõhmus, K.; Majdi, H.; Metcalfe, D.; Olsthoorn, A.F.M.; Pronk, A.; et al. Specific root length as an indicator of environmental change. *Plant Biosyst.* **2007**, *141*, 426–442. [CrossRef]
14. Amendola, C.; Montagnoli, A.; Terzaghi, M.; Trupiano, D.; Oliva, F.; Baronti, S.; Miglietta, F.; Chiatante, D.; Scippa, G.S. Short-term effects of biochar on grapevine fine root dynamics and arbuscular mycorrhizae production. *Agric. Ecosyst. Environ.* **2017**, *239*, 236–245. [CrossRef]
15. Withington, J.M.; Reich, P.B.; Oleksyn, J.; Eissenstat, D.M. Comparisons of structure and life span in roots and leaves among temperate trees. *Ecol. Monogr.* **2006**, *76*, 381–397. [CrossRef]
16. Zadworny, M.; McCormack, M.L.; Rawlik, K.; Jagodziński, A.M. Seasonal variation in chemistry, but not morphology, in roots of Quercus robur growing in different soil types. *Tree Physiol.* **2015**, *35*, 644–652. [CrossRef]
17. Li, A.; Guo, D.; Wang, Z.; Liu, H. Nitrogen and phosphorus allocation in leaves, twigs, and fine roots across 49 temperate, subtropical and tropical tree species: A hierarchical pattern. *Funct. Ecol.* **2010**, *24*, 224–232. [CrossRef]
18. Terzaghi, M.; Montagnoli, A.; Di Iorio, A.; Scippa, G.S.; Chiatante, D. Fine-root carbon and nitrogen concentration of European beech (*Fagus sylvatica L.*) in Italy Prealps: Possible implications of coppice conversion to high forest. *Front. Plant Sci.* **2013**, *4*, 192. [CrossRef]
19. Vajari, K.A.; Jalilvand, H.; Pourmajidian, M.R.; Espahbodi, K.; Moshki, A. The effect of single-tree selection system on soil properties in an oriental beech stand of Hyrcanian forest, north of Iran. *J. For. Res.* **2011**, *22*, 591–596. [CrossRef]
20. Sommerfeld, R.; Lundquist, J.; Smith, J. Characterizing the canopy gap structure of a disturbed forest using the Fourier transform. *For. Ecol. Manag.* **2000**, *128*, 101–108. [CrossRef]
21. Miller, S.D.; Goulden, M.L.; da Rocha, H.R. The effect of canopy gaps on subcanopy ventilation and scalar fluxes in a tropical forest. *Agric. For. Meteorol.* **2007**, *142*, 25–34. [CrossRef]
22. Vepakomma, U.; St-Onge, B.; Kneeshaw, D. Spatially explicit characterization of boreal forest gap dynamics using multi-temporal lidar data. *Remote Sens. Environ.* **2008**, *112*, 2326–2340. [CrossRef]
23. Kukkonen, M.; Rita, H.; Hohnwald, S.; Nygren, A. Treefall gaps of certified, conventionally managed and natural forests as regeneration sites for Neotropical timber trees in northern Honduras. *For. Ecol. Manag.* **2008**, *255*, 2163–2176. [CrossRef]
24. Cannon, J.B.; Brewer, J.S. Effects of Tornado Damage, Prescribed Fire, and Salvage Logging on Natural Oak (Quercus spp.) Regeneration in a Xeric Southern USA Coastal Plain Oak and Pine Forest. *Nat. Areas J.* **2013**, *33*, 39–49. [CrossRef]
25. Izbicki, B.J.; Alexander, H.D.; Paulson, A.K.; Frey, B.R.; McEwan, R.W.; Berry, A.I. Prescribed fire and natural canopy gap disturbances: Impacts on upland oak regeneration. *For. Ecol. Manag.* **2020**, *465*, 118107. [CrossRef]
26. Gagnon, J.L.; Jokela, E.J.; Moser, W.; Huber, D.A. Characteristics of gaps and natural regeneration in mature longleaf pine flatwoods ecosystems. *For. Ecol. Manag.* **2004**, *187*, 373–380. [CrossRef]
27. Yamamoto, S.-I. Forest Gap Dynamics and Tree Regeneration. *J. For. Res.* **2000**, *5*, 223–229. [CrossRef]

28. Schnabel, F.; Donoso, P.J.; Winter, C. Short-term effects of single-tree selection cutting on stand structure and tree species composition in Valdivian rainforests of Chile. *New Zeal. J. For. Sci.* **2017**, *47*, 21. [CrossRef]
29. Muscolo, A.; Bagnato, S.; Sidari, M.; Mercurio, R. A review of the roles of forest canopy gaps. *J. For. Res.* **2014**, *25*, 725–736. [CrossRef]
30. Zhu, J.J.; Tan, H.; Li, F.Q.; Chen, M.; Zhang, J.X. Microclimate regimes following gap formation in a montane secondary forest of eastern Liaoning Province, China. *J. For. Res.* **2007**, *18*, 167–173. [CrossRef]
31. Kooch, Y.; Bayranvand, M. Composition of tree species can mediate spatial variability of C and N cycles in mixed beech forests. *For. Ecol. Manag.* **2017**, *401*, 55–64. [CrossRef]
32. Guner, S.; Yagci, V.; Tilki, F.; Celik, N. The effects of initial planting density on above-and below-ground biomass in a 25-year-old Fagus orientalis Lipsky plantation in Hopa, Turkey. *Sci. Res. Essays* **2010**, *5*, 1856–1860.
33. Brassard, B.W.; Chen, H.Y.H.; Bergeron, Y. Influence of environmental variability on root dynamics in northern forests. *CRC Crit. Rev. Plant Sci.* **2009**, *28*, 179–197. [CrossRef]
34. Pregitzer, K.S.; DeForest, J.L.; Burton, A.J.; Allen, M.F.; Ruess, R.W.; Hendrick, R.L. Fine root architecture of nine North American trees. *Ecol. Monogr.* **2002**, *72*, 293–309. [CrossRef]
35. Guo, D.; Xia, M.; Wei, X.; Chang, W.; Liu, Y.; Wang, Z. Anatomical traits associated with absorption and mycorrhizal colonization are linked to root branch order in twenty-three Chinese temperate tree species. *New Phytol.* **2008**, *180*, 673–683. [CrossRef]
36. Pregitzer, K.S.; Zak, D.R.; Loya, W.M.; Karberg, N.J.; King, J.S.; Burton, A.J. The Contribution of Root-Rhizosphere Interactions to Biogeochemical Cycles in a Changing World. In *The Rhizosphere: An Ecological Perspective*; Cardon, Z.G., Whitbeck, J.L., Eds.; Elsevier Academic Press: Burlington, MA, USA, 2007; pp. 155–178. ISBN 9780120887750.
37. Withington, J.M.; Elkin, A.D.; Bułaj, B.; Olesiński, J.; Tracy, K.N.; Bouma, T.J.; Oleksyn, J.; Anderson, L.J.; Modrzyński, J.; Reich, P.B.; et al. The impact of material used for minirhizotron tubes for root research. *New Phytol.* **2003**, *160*, 533–544. [CrossRef]
38. Long, Y.; Kong, D.; Chen, Z.; Zeng, H. Variation of the linkage of root function with root branch order. *PLoS ONE* **2013**, *8*, e57153. [CrossRef]
39. Iversen, C.M. Using root form to improve our understanding of root function. *New Phytol.* **2014**, *203*, 707–709. [CrossRef]
40. Montagnoli, A.; Terzaghi, M.; Giussani, B.; Scippa, G.S.; Chiatante, D. An integrated method for high-resolution definition of new diameter-based fine root sub-classes of *Fagus sylvatica* L. *Ann. For. Sci.* **2018**, *75*, 76. [CrossRef]
41. Montagnoli, A.; Terzaghi, M.; Di Iorio, A.; Scippa, G.S.; Chiatante, D. Fine-root morphological and growth traits in a Turkey-oak stand in relation to seasonal changes in soil moisture in the Southern Apennines, Italy. *Ecol. Res.* **2012**, *27*, 1015–1025. [CrossRef]
42. Montagnoli, A.; Di Iorio, A.; Terzaghi, M.; Trupiano, D.; Scippa, G.S.; Chiatante, D. Influence of soil temperature and water content on fine-root seasonal growth of European beech natural forest in Southern Alps, Italy. *Eur. J. For. Res.* **2014**, *133*, 957–968. [CrossRef]
43. Montagnoli, A.; Dumroese, R.K.; Terzaghi, M.; Onelli, E.; Scippa, G.S.; Chiatante, D. Seasonality of fine root dynamics and activity of root and shoot vascular cambium in a Quercus ilex L. forest (Italy). *For. Ecol. Manag.* **2019**, *431*, 26–34. [CrossRef]
44. Yoshida, T.; Naito, S.; Nagumo, M.; Hyodo, N.; Inoue, T.; Umegane, H.; Yamazaki, H.; Miya, H.; Nakamura, F. Structural Complexity and Ecosystem Functions in a Natural Mixed Forest under a Single-Tree Selection Silviculture. *Sustainability* **2017**, *9*, 2093. [CrossRef]
45. Alireza, A.; Abrari, V.K.; Mohammad, F.; Antonino, D.I. Influences of forest gaps on soil physico-chemical and biological properties in an oriental beech (*Fagus orientalis* L.) stand of Hyrcanian forest, north of Iran. *iForest* **2020**, *13*, 124–129.
46. Leavitt, S.W.; Danzer, S.R. Method for Batch Processing Small Wood Samples to Holocellulose for Stable-Carbon Isotope Analysis. *Anal. Chem.* **1993**, *65*, 87–89. [CrossRef]
47. Genet, M.; Stokes, A.; Salin, F.; Mickovski, S.B.; Fourcaud, T.; Dumail, J.F.; Van Beek, R. The influence of cellulose content on tensile strength in tree roots. *Plant Soil* **2005**, *278*, 1–9. [CrossRef]
48. Doster, M.A.; Bostock, R.M. Quantification of Lignin Formation in Almond Bark in Response to Wounding and Infection by Phytophthora Species. *Phytopathology* **1988**, *78*, 473. [CrossRef]
49. Trupiano, D.; Di Iorio, A.; Montagnoli, A.; Lasserre, B.; Rocco, M.; Grosso, A.; Scaloni, A.; Marra, M.; Chiatante, D.; Scippa, G.S. Involvement of lignin and hormones in the response of woody poplar taproots to mechanical stress. *Physiol. Plant.* **2012**, *146*, 39–52. [CrossRef]
50. Cambi, M.; Certini, G.; Neri, F.; Marchi, E. The impact of heavy traffic on forest soils: A review. *For. Ecol. Manag.* **2015**, *338*, 124–138. [CrossRef]
51. Gray, A.N.; Spies, T.A.; Easter, M.J. Microclimatic and soil moisture responses to gap formation in coastal Douglas-fir forests. *Can. J. For. Res.* **2002**, *32*, 332–343. [CrossRef]
52. Terzaghi, M.; Di Iorio, A.; Montagnoli, A.; Baesso, B.; Scippa, G.S.; Chiatante, D. Forest canopy reduction stimulates xylem production and lowers carbon concentration in fine roots of European beech. *For. Ecol. Manag.* **2016**, *379*, 81–90. [CrossRef]
53. Ni, X.; Berg, B.; Yang, W.; Li, H.; Liao, S.; Tan, B.; Yue, K.; Xu, Z.; Zhang, L.; Wu, F. Formation of forest gaps accelerates C, N and P release from foliar litter during 4 years of decomposition in an alpine forest. *Biogeochemistry* **2018**, *139*, 321–335. [CrossRef]
54. Scharenbroch, B.C.; Bockheim, J.G. Impacts of forest gaps on soil properties and processes in old growth northern hardwood-hemlock forests. *Plant Soil* **2007**, *294*, 219–233. [CrossRef]
55. Parsons, W.F.J.; Knight, D.H.; Miller, S.L. Root Gap Dynamics in Lodgepole Pine Forest: Nitrogen Transformations in Gaps of Different Size. *Ecol. Appl.* **1994**, *4*, 354–362. [CrossRef]

56. Thomas, F.M.; Molitor, F.; Werner, W. Lignin and cellulose concentrations in roots of Douglas fir and European beech of different diameter classes and soil depths. *Trees—Struct. Funct.* **2014**, *28*, 309–315. [CrossRef]
57. Zhang, C.B.; Chen, L.H.; Jiang, J. Why fine tree roots are stronger than thicker roots: The role of cellulose and lignin in relation to slope stability. *Geomorphology* **2014**, *206*, 196–202. [CrossRef]
58. Zhu, J.; Wang, Y.; Wang, Y.; Mao, Z.; Langendoen, E.J. How does root biodegradation after plant felling change root reinforcement to soil? *Plant Soil* **2020**, *446*, 211–227. [CrossRef]
59. McCormack, M.L.; Iversen, C.M. Physical and Functional Constraints on Viable Belowground Acquisition Strategies. *Front. Plant Sci.* **2019**, *10*, 1–12. [CrossRef]
60. Kong, D.; Wang, J.; Wu, H.; Valverde-Barrantes, O.J.; Wang, R.; Zeng, H.; Kardol, P.; Zhang, H.; Feng, Y. Nonlinearity of root trait relationships and the root economics spectrum. *Nat. Commun.* **2019**, *10*, 1–9. [CrossRef]
61. Kramer-Walter, K.R.; Bellingham, P.J.; Millar, T.R.; Smissen, R.D.; Richardson, S.J.; Laughlin, D.C. Root traits are multidimensional: Specific root length is independent from root tissue density and the plant economic spectrum. *J. Ecol.* **2016**, *104*, 1299–1310. [CrossRef]
62. Fayle, D.C.F. Distribution of radial growth during the development of red pine root systems. *Can. J. For. Res.* **1975**, *5*, 608–625. [CrossRef]
63. Vincent, M.; Krause, C.; Zhang, S.Y. Radial growth response of black spruce roots and stems to commercial thinning in the boreal forest. *Forestry* **2009**, *82*, 557–571. [CrossRef]
64. Talbot, J.M.; Treseder, K.K. Interactions among lignin, cellulose, and nitrogen drive litter chemistry-decay relationships. *Ecology* **2012**, *93*, 345–354. [CrossRef] [PubMed]
65. Walela, C.; Daniel, H.; Wilson, B.; Lockwood, P.; Cowie, A.; Harden, S. The initial lignin: Nitrogen ratio of litter from above and below ground sources strongly and negatively influenced decay rates of slowly decomposing litter carbon pools. *Soil Biol. Biochem.* **2014**, *77*, 268–275. [CrossRef]
66. Zhang, X.; Wang, W. The decomposition of fine and coarse roots: Their global patterns and controlling factors. *Sci. Rep.* **2015**, *5*, 1–10. [CrossRef]
67. Lützow, M.V.; Kögel-Knabner, I.; Ekschmitt, K.; Matzner, E.; Guggenberger, G.; Marschner, B.; Flessa, H. Stabilization of organic matter in temperate soils: Mechanisms and their relevance under different soil conditions—A review. *Eur. J. Soil Sci.* **2006**, *57*, 426–445. [CrossRef]
68. Valverde-Barrantes, O.J.; Smemo, K.A.; Blackwood, C.B. Fine root morphology is phylogenetically structured, but nitrogen is related to the plant economics spectrum in temperate trees. *Funct. Ecol.* **2015**, *29*, 796–807. [CrossRef]
69. Comas, L.H.; Eissenstat, D.M.; Lakso, A.N. Assessing root death and root system dynamics in a study of grape canopy pruning. *New Phytol.* **2000**, *147*, 171–178. [CrossRef]
70. Lee, M.H.; Comas, L.H.; Callahan, H.S. Experimentally reduced root-microbe interactions reveal limited plasticity in functional root traits in Acer and Quercus. *Ann. Bot.* **2014**, *113*, 513–521. [CrossRef]
71. Comas, L.H.; Eissenstat, D.M. Patterns in root trait variation among 25 co-existing North American forest species. *New Phytol.* **2009**, *182*, 919–928. [CrossRef]

Article

Evaluating Soil–Root Interaction of Hybrid Larch Seedlings Planted under Soil Compaction and Nitrogen Loading

Tetsuto Sugai [1], Satoko Yokoyama [2], Yutaka Tamai [3], Hirotaka Mori [3], Enrico Marchi [4], Toshihiro Watanabe [1], Fuyuki Satoh [5] and Takayoshi Koike [1,2,*]

1 Plant Nutrition laboratory, Hokkaido University, Sapporo 060-8689, Japan; tsugai@for.agr.hokudai.ac.jp (T.S.); nabe@chem.agr.hokudai.ac.jp (T.W.)

2 Silviculture and Forest Ecological Studies, Hokkaido University, Sapporo 060-8589, Japan; s-yokoyama@frontier.hokudai.ac.jp

3 Forest Bioresource Technology laboratory, Hokkaido University, Sapporo 060-8589, Hokkaido, Japan; ytamai@for.agr.hokudai.ac.jp (Y.T.); mrh28m@eis.hokudai.ac.jp (H.M.)

4 Department of Agriculture, Food, environment and Forestry, University of Florence, Via S. Bonaventura 13, 50145 Firenze, Italy; enrico.marchi@unifi.it

5 Field Science Center for Northern Biosphere, Hokkaido University, Sapporo 060-8589, Hokkaido, Japan; f-satoh@fsc.hokudai.ac.jp

* Correspondence: tkoike@for.agr.hokudai.ac.jp

Received: 22 June 2020; Accepted: 25 August 2020; Published: 29 August 2020

Abstract: Although compacted soil can be recovered through root development of planted seedlings, the relationship between root morphologies and soil physical properties remain unclear. We investigated the impacts of soil compaction on planted hybrid larch F_1 (*Larix gmelinii* var. *japonica* × *L. kaempferi*, hereafter F_1) seedlings with/without N loading. We assumed that N loading might increase the fine root proportion of F_1 seedlings under soil compaction, resulting in less effects of root development on soil recovery. We established experimental site with different levels of soil compaction and N loading, where two-year-old F_1 seedlings were planted. We used a hardness change index (HCI) to quantify a degree of soil hardness change at each depth. We evaluated root morphological responses to soil compaction and N loading, focusing on ectomycorrhizal symbiosis. High soil hardness reduced the total dry mass of F_1 seedlings by more than 30%. Significant positive correlations were found between HCI and root proportion, which indicated that F_1 seedling could enhance soil recovery via root development. The reduction of fine root density and its proportion due to soil compaction was observed, while these responses were contrasting under N loading. Nevertheless, the relationships between HCI and root proportion were not changed by N loading. The relative abundance of the larch-specific ectomycorrhizal fungi under soil compaction was increased by N loading. We concluded that the root development of F_1 seedling accelerates soil recovery, where N loading could induce root morphological changes under soil compaction, resulting in the persistent relationship between root development and soil recovery.

Keywords: soil compaction; N loading; fine root; root morphology; ectomycorrhizal fungi

---

1. Introduction

Impacts of forest soil compaction that are caused by forestry machine operation have serious concerns for sustainable forest management [1–3]. Soil compaction degrades forest ecosystem functions [4], such as nutrient availability [5,6], the community of soil macro- and microorganism [7], and tree growth [8–10]. High soil hardness can inhibit root development, resulting in the inhibition of physiological activities and growth suppression of trees [4]. Soil hardness means pressure resistance and it is a general physical parameter [11–13]. The effects of soil compaction on tree growth vary depending on soil type, condition, and tree species [14–16]. Manipulation experiments can contribute to more accurate understanding of tree growth response to soil compaction and its mechanisms [17–19].

The recovery of soil structures and functions from compaction (hereafter, soil recovery) varies depending on several site-related factors, such as soil texture, organic matter content, pedoclimate, and biomass and activity of soil biota [20–23]. Soil recovery can be associated with the root growth of planted tree seedlings, which has been primarily studied in alder species (*Alnus* sp.) [24,25], although the alder planting has raised other ecological concerns in forest management [26]. Root development can penetrate the compacted soil layer [27] and form cracks and fissures in soil [28,29]. Root development under compacted soil would be affected by soil biotic and abiotic factors [30]. Although fertilizer application could promote both aboveground and belowground growth, even in compacted soil [31,32], few studies have elucidated root morphological responses and the symbiosis with soil microorganisms under soil compaction and N loading [33]. Because artificial construction to improve compacted soil properties has economic and ecological issues [6,24], it is of outmost importance to investigate the relationship between roots of planted seedlings and soil physical properties for the ecological managements of forest soil [1].

Larch (*Larix* sp.) trees have widely been planted for timber production in Northeast Asia [34,35]. Among larch species, hybrid larch F_1 (hereafter F_1) has been developed for the superior initial growth [34,36]. However, the overuse of forestry machines has been concerned in afforestation regions [37–39]. The effects of N loading on tree growth and the symbiosis between ectomycorrhizal (ECM) fungi have been extensively investigated [40,41]. N loading promotes the growth of F_1 seedling [42], whereas excess N loading would reduce its stress resistance [43]. A stable ECM community of F_1 seedlings was observed under N loading [41]. Therefore, we supposed that F_1 would be a suitable model tree species to verify the responses of root development to soil compaction and N loading under the consistent relationship with ECM fungi.

In this study, we investigated the growth of F_1 seedlings planted in the compacted soil where N was added. Given the root response to soil physical [31] and nutritional conditions [43], we hypothesized that N loading would increase the fine root proportion of F_1 seedlings under soil compaction, resulting in a reduced effect of root development on soil recovery. We attempted to verify the following objectives: (1) whether the root developments of F_1 would affect soil hardness and (2) how the morphological traits respond to soil compaction and N loading. We synthesized these results to reveal the capacity of promoting soil recovery by planting seedlings for forest ecological management [44].

2. Material and Methods

2.1. Site and Experimental Design

This experiment was conducted in Sapporo Experimental Forest of Hokkaido University, Japan (43°04′ N, 141°20′ E, 15 m a.s.l.), during two growing seasons from May 2018 to November 2019. The climate is cool temperate with high humidity. In 2018, the mean annual air temperature was 9.1 °C and the total precipitation was 1282 mm and, in 2019, the mean annual air temperature was 9.5 °C and the total precipitation was 814 mm. The snow-free period is mid-April to early November. Meteorological data were recorded by the Japan Meteorological Agency near the study site. The experimental plots were established at the nursery under full sunlight to make them as uniform as possible. The soil of the experimental area was classified as brown forest soil (Dystric Cambisols) [45].

We examined the effects of soil compaction and N loading using a split-plot design (Figure S1). Six combinations were set with five replicate plots for each treatment (three levels of soil compaction × 2 N loading levels × 5 replicated blocks). First, a 40 m track line for each compaction level was divided into five blocks. The distance from each track line was 1 m, with a trench of 40 cm depth. Each block size was approximately 1.5 m × 6.0 m, and each block was separated by at least 1 m.

Each block was divided into two subplots for N loading treatment (a total of 30 subplots). A subplot size was approximately set as 1.5 m × 2.5 m with a distance of 1 m from each other.

One-year-old seedlings of hybrid larch F_1 (*L. gmelinii* var. *japonica* × *L. kaempferi*) were produced at a nursery of Hokkaido Research Organization, Forestry Research Institute. Before needle unfolding, four seedlings were transplanted into all subplots on 15 May 2018 (a total of 120 seedlings). A seedling was planted into the space excavated (0.1 m × 0.1 m × 0.4 m), and the distance from each seedling was at least 0.5 m in a subplot. The initial size (means ± SD) was 25.53 ± 2.17 cm for aboveground height and 6.00 ± 0.87 mm for collar diameter. After planting, germinated grasses that were around the F_1 seedlings were manually removed.

A root tillage treatment was applied by a cultivator for mixing the soil to a depth of approximately 30 cm in order to set high environmental homogeneity of the study site. Weeding was carried out before the soil compaction treatment. The soil was compacted with heavy machine passes conducted on 11 May 2018. The target levels of surface soil hardness were set as 4.7 and 14.0 kg cm^{-2} based on the national environmental standards [46]. The heavy machines passed over the study sites until the target of the soil compaction level was achieved, as measured using a soil hardness tester (Yamanaka's Soil Hardness tester, Fujiwara Scientific Co., Ltd., Tokyo, Japan). Two types of heavy machinery were operated on 1.5 m wide and 40 m long tracks: a 1.0 t mini excavator (U-008, KUBOTA, Osaka, Japan) was used for the low compaction level and a 2.0 t tractor (TA278F, ISEKI, Matsuyama, Japan) was used for the high compaction level. Three levels of soil compaction were simulated: control, undisturbed soil after tillage; low compaction treatment obtained with 10 passes of a mini excavator after tillage; and, high compaction obtained with 23 passes of a tractor after tillage.

Two levels of N loading were simulated: 0 (−N) and 50 kg N ha^{-1} $year^{-1}$ (+N). The N loading level was reflected with the observed maximum value in Japan [47]. Ammonium sulfate ($(NH_4)_2SO_4$) was used because of concerns of the recent N deposition in Japan [48]. In plots of +N, N loading was conducted on 6 July, 20 July, 17 August, 25 September, and 18 October 2018, without rainy days. $(NH_4)_2SO_4$ was dissolved in tap water and then applied directly on soil, avoiding the application on seedling bodies. The applied volume for a seedling was approximately 500 mL during each application. Plots of −N were irrigated with the same amount of tap water on the treatment dates.

2.2. Soil Environment

In this study, soil hardness (SH) was defined as the pressure resistance acquired by a conical tip probe pushing into soil using a soil hardness tester (Yamanaka's Soil Hardness tester, Fujiwara Scientific Co., Ltd., Tokyo, Japan). Soil bulk density (SBD) is an indicator of soil packing degree and it is calculated as the dry weight per unit volume of soil.

The vertical SH was measured along with four depths: 0–10, 10–20, 20–30, and 30–40 cm. These measurements were taken during the first and final research periods. The first SH along with depth were measured on 15 May 2018. A mini pit was manually dug into soil with a volume of approximately 10 cm × 35 cm × 45 cm. The position of the hole was randomly set in each subplot at least 30 cm away from the planted seedling. Subsequently, SH was randomly measured in a soil face at each depth. The mean value of these measurements was used as a representative SH value for each seedling. The final SH along with depth was measured in a space with a volume of approximately 50 cm × 50 cm × 40 cm from the end of October to early November in 2019, where a seedling was excavated. The three points where the SH was measured were randomly selected at each depth after excavating a seedling, and the mean value was calculated for each seedling. The hardness change index (hereafter HCI) was calculated as the difference between the final and first mean values of SH to evaluate the relationship between soil variation and root development considering soil depth.

The surface SH was also measured on 15 May, 5 June, 20 July, 11 September, and 23 October 2018, and on 1 May, 3 September, and 2 November 2019. The surface SH was measured at four to six random points approximately 20 cm away from the seedling stem. The mean value of the points was used as a representative SH value for each subplot.

Surface soil sampling was conducted in order to measure SBD and soil water contents (SWC) on 5 June, 20 July, 11 September, and 23 October 2018, avoiding the rainy days. Soil was sampled at each subplot using a metal cylinder (5 cm × 20 cm^2, Daiki Rika Kogyo Co., Ltd., Saitama, Japan) without any plant litter. After sampling, the fresh soil mass was immediately weighed using a 0.1 g scale (EB-3300SW, Shimadzu Co., Ltd., Kyoto, Japan) and then dried for 24 h at 105 °C.

On 15 August 2018, another soil sampling was conducted at each subplot in order to determine three phase fractions of soil. Solid and liquid phases were measured using a digital volume analyzer (DIK-1150, Daiki Rika Kogyo Co., Ltd., Saitama, Japan), and the residual was calculated as the air phase fraction.

Fresh soil was sampled for its chemical properties on August 2018. The sampled soil was passed through a stainless-steel sieve with a 1.0 mm mesh. Subsequently, 10 g of soil was mixed with 25 mL of 2 M potassium chloride solution. The samples were shaken for over 1 h before measurement using a pH sensor (RM-30P, TOA DKK, Co., Ltd., Tokyo, Japan). The soil inorganic N contents of the same samples were measured. The samples were filtered through a 1-μm filter (No. 5C filter paper, Toyo Roshi, Tokyo, Japan) for the colorimetric measurement of inorganic N (total value of NH_4^+–N and NO_3^-–N) using a flow injection analyzer (AQLA-700, Aqualab Co., Ltd., Tokyo, Japan).

2.3. Seedling Performances

2.3.1. Aboveground

The height and the collar diameter of 120 seedlings were regularly measured using a measuring tape with 0.1 cm resolution and a digital caliper with 0.01 mm resolution (Mitsutoyo, Kanagawa, Japan) on 10 June, 6 July, 7 August, 5 September, and 8 October 2018, and 14 October 2019. The diameter was calculated as the mean of two crosswise measurements. The xylem pressure potential of the shoot tips was measured on 9 September 2019. Five seedlings were randomly selected for each treatment, and the sampled intact shoot was measured using a pressure chamber (PMS-600, PMS Instrument Co., Albany, OR, USA) at predawn (03:30–04:30 h) and mid-day (13:00–14:00 h).

On 21 October 2019, the aboveground plant body of 120 seedlings was harvested. The harvested biomass of each seedling was divided by stems, branches, and needles. The stems were divided into three sections at the internode: a current-year stem grown in 2019, a one-year stem formed in 2018, and the two-year stem. The height and collar diameter of these stems were measured using the same methods described above. All of the branches with needles were divided into three sections, following the same approach used for the stems. The branches of each stem were counted. The longest branch was sampled in a current-year stem, and its length and bottom diameter were measured. Subsequently, all of the separated organs were put into a dry oven at 80 °C for two weeks to determine the dry mass. We defined the branch productivity index (BPI) and needle productivity index (NPI) as the following formulas in order to evaluate the morphological balance and potential productivity in a current-year stem:

$$BPI\left(\mathrm{m}^{-1}\right) = \frac{\mathit{The\ number\ of\ branch\ in\ current\text{–}year\ stem}}{\mathit{The\ length\ of\ current\ stem}\ (\mathrm{m})}, \quad (1)$$

$$NPI\left(\mathrm{g\ m}^{-1}\right) = \frac{\mathit{Needle\ dry\ weight\ in\ current\text{–}year\ stem}\ (\mathrm{g})}{\mathit{The\ length\ of\ maximum\ current\text{–}year\ branch}\ (\mathrm{m})}. \quad (2)$$

2.3.2. Belowground

The root systems of the seedlings were carefully excavated at the control and at a high level of soil compaction treatment from the end of October to early November in 2019. The root systems were sampled for three seedlings of each subplot (two compaction levels × 2 N loading levels × 5 replicated blocks × 3 biological replicates, a total of 60 seedlings). Before sampling, the excavating area and depth were set to approximately 50 cm × 50 cm × 40 cm. The excavated volume of space was measured using a measuring tape with 1 cm resolution just after root sampling to avoid any soil deformation.

The root system was carefully washed in order to eliminate soil and separated visually into a rootstock and other roots. The other roots were divided into lateral roots (diameter ≥ 2 mm) and fine roots (diameter < 2 mm) using a digital caliper. Finally, the separated roots were dried in an oven at 65 °C for a week, and the dry weight was measured. These weights were used to calculate root density as the ratio between the root weight and excavated soil volume.

When excavating root systems of seedlings, a seedling was randomly selected in each subplot in order to determine the fine root morphological traits and the ECM association (two compaction levels × 2 N loading levels × 5 replicated blocks × 1 biological replicates, a total of 20 seedlings). Approximately six fine roots were collected along each distance from stem (0–10, 10–20, and 20–30 cm). The sampled roots were covered with wet papers, stored in a plastic bag, and transferred to a dark refrigerator (4 °C) in the laboratory until the following measurements. The remaining soil was carefully removed from fine roots using a brush. Three intact fine roots were randomly selected from each seedling and scanned using a double-lamp bed scanner with 800 dpi (GT-X 970, Epson, Japan). The scanned image was analyzed to determine the total length, surface area, and volume of each fine root while using root analysis software WinRHIZOTRON 2012 (Regent Instruments, Quebec, QC, Canada). After scanning, the fine roots were dried in an oven at 65 °C for a week, and the dry weight was measured. These values were used in calculating the following morphological trait values as a ratio between the acquired values from the image analysis and the dry weight: specific root length (SRL, a ratio between the total length and dry weight), specific root area (SRA, a ratio between the total surface area and dry weight), and root tissue density (RTD, a ratio between the total volume and dry weight). The remaining sampled fine roots that were collected by each distance were pooled in each treatment and used to evaluate the association of ECM. We classified the ECM taxa using the morphological characteristics that were observed with a microscope (Olympus szx-ILLK100, Tokyo, Japan). The relative abundance of ECM was calculated as the ratio of the number of ECM root tips to the non-ECM root tips. The ECM taxa were verified based on the methods of a previous study by Wang et al. [41].

2.4. Data Analysis

MS EXCEL 2010 (Microsoft ©) and R version 3.4.6 [49] were used for data processing and statistical analysis. Each response variable was calculated as the representative values per experimental block (five values for each treatment for statistical analysis). We evaluated the effects of the compaction and N loading by ANOVA with chi-squared test. ANOVA with chi-squared test was performed on SRL, SRA, and RTD between the horizontal distance from the root trunk and soil compaction at each N treatment. We analyzed Pearson's correlation coefficient between the three types of root density (total root density, lateral root density, and fine root density), SH of the first and final conditions, and HCI. The values in each N loading treatment were pooled for the correlation analysis at first in order to evaluate whether the relationships between the soil and root vary depending on compaction. A significant relationship was identified using the Bonferroni test. In the detected significant relationships, we analyzed whether the effects of the root traits on the soil properties vary by N loading using ANCOVA as a post-hoc test.

3. Results

3.1. Environment at the Study Site

Figure 1 shows the seasonal variations of the meteorological factor and surface SH. The total precipitation from May to November 2018 was 828 mm, and that from May to October 2019 was 482 mm. The compacted surface soil clearly showed a decline in hardness in 2018, whereas the variation was relatively small in 2019 (Figure 1). In 2018, the compaction treatments reduced the air fraction by more than 35% and increased bulk density by more than 50% (Table S1), whereas N loading increased the total inorganic N contents by more than 20%. Therefore, the soil environment was successfully manipulated for the experimental design. Table S2 shows the variations of SH along with soil depth. A negative variation in SH in 2018 was found for high compaction plots: from 1.57 MPa for the surface soil layer to 0.51 MPa for the deepest layer. In 2019, the variation was positive at the control plots, and a negative variation was not observed at the high compaction plots.

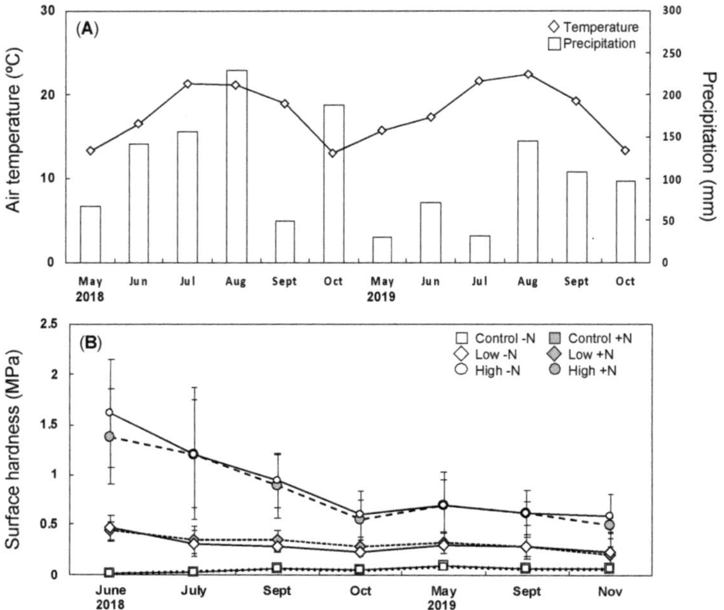

Figure 1. Meteorological and soil conditions at the study area. (**A**) The meteorological conditions in each month; Diamond: mean air temperature; Bar: total precipitation. (**B**) The seasonal changes in average surface SH ± standard error (SE); square: control; diamond: low compaction level; circle: high compaction level; white: non-N loading (−N); gray: N loading (+N). ($n = 5$).

3.2. Aboveground Responses

The effects of soil compaction and N loading were significant on most aboveground traits (Table 1). The final size of the seedlings was reduced in high compaction treatment by approximately 10.2% and 23.1% for the total height and collar diameter, respectively. Compaction significantly reduced the total dry mass by more than 30%. The interaction effects of compaction and N loading were observed ($p < 0.05$). The increased effects of N loading on dry mass were only significant at the control and not on compaction treatments. BPI was reduced by only the high compaction level under N loading, whereas NPI was significantly reduced by compaction alone ($p < 0.05$). A significant reduction of xylem water potential was observed in compacted soil at predawn (Table S3).

Table 1. Mean value ± SE of aboveground trait responses to soil compaction without N loading (−N) and with N loading (+N) and results of ANOVA (Compaction: C; N loading: N). BPI: branch productivity index; NPI: needle productivity index; ***: $p < 0.001$; **: $p < 0.01$; *: $p < 0.05$; N.S.: not significant. ($n = 5$).

		Control		Low Compaction		High Compaction		ANOVA		
Aboveground Traits		−N	+N	−N	+N	−N	+N	C	N	C × N
Stem (cm)	Current	157.59 ± 2.91	164.7 ± 6.08	160.09 ± 3.67	152.92 ± 5.97	151.25 ± 3.64	140.93 ± 8.05	*	N.S.	N.S.
	1-year	42.64 ± 1.50	46.2 ± 2.13	43.94 ± 2.28	45.32 ± 2.39	35.91 ± 3.68	36.82 ± 4.54	**	N.S.	N.S.
	2-year	20.27 ± 0.75	20.53 ± 1.17	18.80 ± 2.18	20.14 ± 0.48	20.04 ± 0.74	20.29 ± 0.74	N.S.	N.S.	N.S.
Diameter (mm)	Current	16.47 ± 1.89	19.13 ± 2.10	16.96 ± 2.28	16.71 ± 2.35	15.90 ± 1.66	14.21 ± 2.82	***	N.S.	**
	1-year	18.33 ± 6.20	22.97 ± 2.10	20.60 ± 2.09	20.10 ± 2.15	18.17 ± 2.91	16.68 ± 4.03	**	N.S.	**
	Collar	30.33 ± 2.85	34.45 ± 2.59	29.96 ± 3.44	29.75 ± 3.33	26.52 ± 3.07	23.34 ± 5.12	***	N.S.	**
Total (g)		281.54 ± 16.89	386.49 ± 25.86	275.11 ± 14.70	277.53 ± 20.32	225.02 ± 17.04	213.15 ± 24.24	***	N.S.	*
Stem (g)	All	166.56 ± 8.26	225.01 ± 14.25	171.18 ± 9.97	169.08 ± 10.48	135.86 ± 10.08	130.92 ± 12.14	***	N.S.	*
	Current	56.27 ± 4.50	79.81 ± 6.79	61.56 ± 4.41	57.67 ± 4.46	50.51 ± 3.62	45.95 ± 4.36	**	N.S.	*
Branch (g)	All	51.27 ± 5.11	73.91 ± 6.44	46.78 ± 2.81	51.16 ± 5.46	38.29 ± 3.19	33.45 ± 5.41	***	*	*
	Current	13.26 ± 2.15	25.52 ± 4.34	15.14 ± 1.70	14.62 ± 2.92	17.61 ± 1.81	11.61 ± 2.41	N.S.	N.S.	*
Needle (g)	All	63.70 ± 5.54	87.56 ± 7.27	57.13 ± 3.63	57.28 ± 5.90	50.86 ± 4.90	48.77 ± 8.05	***	N.S.	N.S.
	Current	50.53 ± 4.48	72.16 ± 6.52	47.11 ± 3.08	46.17 ± 4.94	43.61 ± 4.26	40.78 ± 6.43	**	N.S.	N.S.
Current stem	BPI (m^{-1})	16.18 ± 1.88	18.51 ± 1.68	16.54 ± 1.68	15.70 ± 1.61	17.82 ± 1.80	10.27 ± 1.32	N.S.	N.S.	*
	Length (cm)	67.05 ± 10.88	78.31 ± 18.05	92.24 ± 23.31	90.14 ± 26.95	69.31 ± 11.46	37.13 ± 3.22	N.S.	N.S.	N.S.
Maximum branch	Diameter (mm)	4.77 ± 0.32	5.01 ± 0.38	4.47 ± 0.20	4.26 ± 0.43	4.44 ± 0.26	3.64 ± 0.20	N.S.	N.S.	N.S.
	Total (g)	52.59 ± 4.64	74.28 ± 6.76	48.92 ± 2.96	47.89 ± 5.05	45.52 ± 4.51	41.72 ± 6.46	**	N.S.	N.S.
	Needle (g)	2.06 ± 0.39	2.12 ± 0.44	1.80 ± 0.24	1.72 ± 0.31	1.91 ± 0.37	0.93 ± 0.08	N.S.	N.S.	N.S.
	NPI (g m^{-1})	3.00 ± 0.20	3.11 ± 0.32	2.51 ± 0.17	2.36 ± 0.22	2.61 ± 0.17	2.50 ± 0.09	*	N.S.	N.S.

3.3. Belowground Responses

The effects of soil compaction were significant on all root densities (Table 2). Significant effects of N loading were not observed in lateral and fine root densities. Without N loading, the reduction of root density was approximately 50% in lateral and fine root, but less than 30% in the rootstock. The interaction effects of soil compaction and N loading were observed in both the rootstock ($p < 0.05$) and fine root density ($p < 0.001$). The effect of soil compaction was almost doubled under N loading in rootstock, but the fine root density under N loading was not changed by soil compaction. The lateral root proportion was significantly decreased by soil compaction ($p < 0.05$). The fine root proportion did not change significantly under soil compaction. However, the interaction effect of soil compaction and N loading was observed ($p < 0.05$). The reduction of fine root proportion due to soil compaction was not observed under N loading, rather its proportion was increased. The interaction effect of soil compaction and N loading was also observed in rootstock ($p < 0.05$), showing that N loading only increased its proportion without soil compaction.

Table 2. Mean value ± SE of underground responses to soil compaction without N loading (−N) and with N loading (+N) and the results of ANOVA (Compaction: C; N loading: N). ***: $p < 0.001$; *: $p < 0.05$; N.S.: not significant. ($n = 5$).

		Control		High Compaction		ANOVA		
Traits	Root Type	−N	+N	−N	+N	C	N	C × N
Root density ($kg\ m^{-3}$)	Total	0.90 ± 0.08	1.16 ± 0.18	0.54 ± 0.07	0.68 ± 0.08	***	***	N.S.
	Rootstock	0.37 ± 0.06	0.60 ± 0.20	0.28 ± 0.09	0.33 ± 0.03	***	***	*
	Lateral	0.41 ± 0.07	0.46 ± 0.09	0.2 ± 0.02	0.26 ± 0.06	***	N.S.	N.S.
	Fine	0.12 ± 0.02	0.09 ± 0.02	0.06 ± 0.01	0.08 ± 0.02	***	N.S.	***
Root proportion to total root biomass (%)	Rootstock	41.02 ± 6.27	51.9 ± 3.31	50.79 ± 4.78	49.32 ± 5.41	N.S.	N.S.	*
	Lateral	13.6 ± 2.82	8.47 ± 3.12	11.29 ± 2.63	12.27 ± 1.96	*	N.S.	N.S.
	Fine	45.38 ± 5.08	39.63 ± 3.03	37.92 ± 2.45	38.41 ± 4.34	N.S.	N.S.	*

3.4. Relationship between Soil Hardness and Roots

Correlation analysis showed several significant relationships between SH and root density (Table S4). The highest correlation was observed between the fine root proportion and surface HCI under compaction treatment (Figure 2B, $p < 0.001$). The same positive correlation was observed in the lateral root proportion under compaction treatment (Figure 2A, $p < 0.01$). The second highest correlation was between the 10–20 cm of HCI and total root density of the control ($p < 0.001$). The relationship between the 10–20 cm of HCI and fine root proportion was negative ($p < 0.001$). Similar tendencies of these relationships were not observed at the control or the compaction treatment (Table S4). The results of the post-hoc test showed that there was no significant effect of N loading on the relationship between HCI and each root proportion (Table S5).

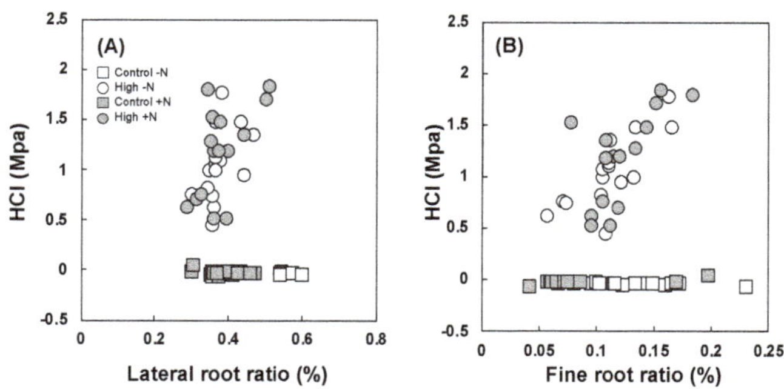

Figure 2. Relationships between (**A**) lateral root proportion and (**B**) fine root proportion and HCI ($n = 5$). Circle: control; square: high compaction; white: no N loading (−N); gray: N loading (+N).

3.5. Responses of Root Morphological Traits and ECM Association

SRL was significantly reduced along with the horizontal distance from the root trunk without N loading, regardless of soil compaction ($p < 0.05$, Table 3). The maximum reduction of SRL was approximately 40% (Figure 3A). By contrast, there were significant interaction effects of distance and compaction under N loading ($p < 0.001$), indicating that the root under N loading would develop depending on the SH. The maximum increment of SRL was nearly double under N loading (Figure 3B). The same pattern was observed in SRL, although there was no significant effect of distance without N loading. Moreover, RTD was significantly increased by soil compaction ($p < 0.05$, Table 3), where the average increment was approximately 20% (Figure 3E).

Table 3. Summary of ANOVA for responses of root morphological traits, specific root length (SRL), specific root area (SRA), and root tissue density. (RTD) without N loading (−N) and with N loading (+N). ***: $p < 0.001$; *: $p < 0.05$; N.S.: not significant. ($n = 5$).

		SRL		SRA		RTD	
N Loading	**Variable**	χ^2	***p* value**	χ^2	***p* Value**	χ^2	***p* Value**
−N	Distance (D)	4.23	*	1.79	N.S.	0.04	N.S.
	Compaction (C)	0.06	N.S.	0.51	N.S.	3.90	*
	D × C	0.27	N.S.	0.08	N.S.	0.03	N.S.
+N	D	0.27	N.S.	0.67	N.S.	0.85	N.S.
	C	1.92	N.S.	0.38	N.S.	0.29	N.S.
	D × C	11.75	***	5.18	*	0.74	N.S.

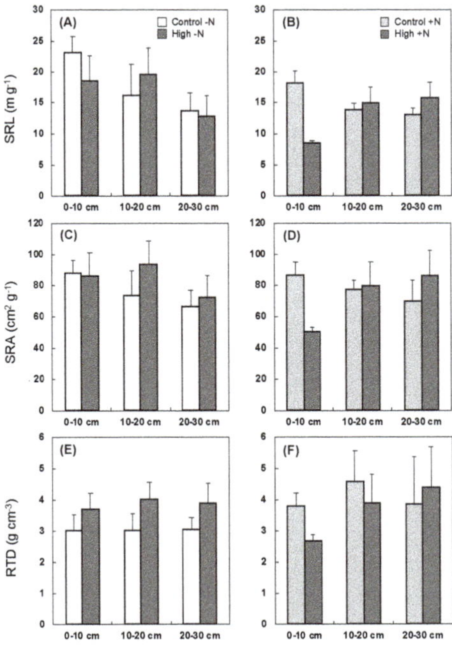

Figure 3. Mean value ± SE in specific root length (SRL), specific root area (SRA), and root tissue density (RTD) ($n = 5$). The responses were showed without N loading (**A**,**C**,**E**) and with N loading (**B**,**D**,**F**). White and light gray: no compaction; dark gray: high compaction.

The ECM taxa were composed of 55 organisms that were identified to species levels, 23 identified to genus level, and 18 as anonymous fungi (Table S6). The most abundant species was *Wilcoxina*, except for the soil compaction and N loading sites (Table 4). The genus *Suilus*, a larch-specific ECM, was the second most abundant, especially at soil compaction and N loading sites.

Table 4. Results of relative abundance (%) for top five in associated ECM under soil compaction and N loading (n = 5). The identified fungal taxa were described as genus level.

Control				High Compaction			
−N		+N		−N		+N	
Wilcoxina	33.87	*Wilcoxina*	31.42	*Wilcoxina*	34.07	*Suillus*	42.73
Suillus	23.44	*Suillus*	19.14	*Suillus*	16.31	*Wilcoxina*	15.86
Laccaria	10.10	*Dothideomycetes*	17.06	*Inocybe*	11.82	*Tetracladium*	10.77
Helicorhoidion	5.41	*Epicoccum*	8.81	*Tomentella*	8.34	*Saitozyma*	6.32
Articulospora	3.46	*Tomentella*	5.14	*Tricholoma*	7.08	*Cryptococcus*	3.45

4. Discussion

4.1. Relationship between Roots and Soil

Significant effects of root development on soil physical parameters were found under soil compaction treatments (Figure 2, Table S4). Namely, the root developments of F_1 seedlings could enhance the recovery of compacted soil, which was supported by previous studies [28–30]. However, in contrast to the results that were found with alder species [30], almost no effects were found in soil condition at a deeper layer (Table S4). Our results would be associated with the shallow root system of larch [34]. Dahurian larch, as one parent of F_1, would develop a shallow root system as an adaptation to a thinner soil layer in the presence of permafrost [34,35]. In fact, there were hardly any roots deeper than the excavating depth of 40 cm when root systems were dug up. Rather, the root systems under soil compaction stress showed the shrunken shape. These results indicated that the effects of F_1 root developments on soil recovery would mainly occur in shallow soil layers. Further, the effects of N loading were not significant on the relationship between HCI and root traits (Table S5), which could be supported by the persistent relationship between roots and soil physical parameters under N application [50]. Soil recovery induced by root development of planted F_1 might not be associated with soil N condition, even though N loading increased lateral and fine root proportions [42,43]. Possible reasons for the persistent relationship between root development and soil recovery could be linked to more specific root morphological responses to soil compaction and N loading.

Relatively shallow soil layers could regenerate faster than the deeper layer [51] because of the soil surface-specific events, such as swelling, shrinking, and freezing [1,6,20]. These studies supported our results that the seasonal physical changes of surface soil were significant (Figure 1). Indeed, the significant physical changes of surface soil in the first year may be partly due to the relatively high amount of precipitation [1,8,20]. However, there were non-significant effects of N loading on most physical soil parameters (Tables S1 and S2). These results would be supported by previous studies [51]. The effects of chemical treatment on forest soil have been mainly evaluated from the lime application [52,53], and little is known regarding N application [31,32]. Further, soil physical dynamics could be affected by biological and chemical factors, like soil macro- and micro-fauna [7,28]. Future studies are needed in order to elucidate the effects of N addition on the physical properties of forest soil, focusing on the interaction between soil chemical and biological factors [22].

4.2. Root Morphological Developments and ECM Symbiosis

It has been reported that there are two types of root development under soil compaction [29]: (i) an avoidance type with higher root developments for allowing the resource exploration in compacted soils and (ii) a resistance type with lower root morphological traits for improving root growth per acquired resources, such as water and nutrients. Non-significant changes in SRL and SRA without N loading (Figure 3) indicated that the studied F_1 seedlings might be a resistance type that is associated with lower plasticity in root development [29]. However, contrasting responses occurred with N loading. That is, root development under soil compaction would vary depending on N loading, which means that the physical response capacity of the root would be altered by soil chemical factors. The morphological response at a specific root level can be interpreted as a strategy for efficient resource acquisition [29,54]. We assumed that F_1 seedlings would acclimate to soil compaction via both the carbon allocation changes within the root system and specific root morphological changes, resulting in a persistent relationship between root development and soil recovery. Moreover, the diversity of ECM-associated F_1 seedlings under soil compaction was changed by N loading (Table 4). The ECM community may vary between different soil nutritional conditions, i.e., N and P [40]. In particular, the specific symbiotic relationship with *Suillus* sp. was reported in larch species [41], which may accelerate the root development of host plants [55]. The shorter branching fine roots of F_1 seedlings may be associated with the specific ECM community under soil compaction and N loading given that ECM fungi could induce a higher frequency of root tips. Because the effects of association and treatments could not be separated in this study, further investigations are required to more accurately understand the root–ECM interaction. In the future, the quantitative effects of root morphological traits on soil recovery also need to be evaluated [1,24,56].

4.3. Aboveground Responses

Different response patterns between NPI and BPI without the N loading condition indicated that BPI would be persistent more than NPI under compacted soil. Branch number is mostly determined by the number of buds and its flush condition [57]. Thus, the persistent BPI indicated that soil compaction would not suppress shoot branching. It has been reported that soil compaction stimulated the regulation of aboveground growth via phytohormonal [4,58]. The plastic responses of shoot branches to SH may be a cue for investigating the interaction between above- and below-ground growth with the complex hormonal regulations. The lower NPI under soil compaction indicated that soil compaction would inhibit the leaf production and growth in F_1 via water stress. Given the lower xylem water potential (Table S3), the photosynthesis of F_1 seedlings would be suppressed by soil compaction, resulting in NPI reduction.

5. Conclusions

Significant positive correlations between both fine and lateral root proportion and HCI indicated that the root development of F_1 seedlings was associated with the soil recovery after compaction. N loading promoted root development, especially under soil compaction. N loading increased fine root density and its proportion under soil compaction, which might be associated with high susceptibility to drought [43]. Indeed, a significant interaction effect of soil compaction and N loading was observed on current-year growth parameters (Table 1). The effects of lower precipitation and soil moisture condition during the growing period in 2019 were also significant (Figure 1). As the water condition was not manipulated in this study, future studies are needed in order to verify the responses to drought tolerance under soil compaction and N loading, where the relationship between above-, and belowground responses should be linked. N loading also changed specific root morphological traits, which could interact with ECM symbiosis, resulting in the retention of the relationship between SH and root traits, even with high proportions of fine roots. Our study addressed the understanding of the capacity of planted F_1 seedlings for soil recovery in ecological silviculture management.

Supplementary Materials: The following are available online at http://www.mdpi.com/1999-4907/11/9/947/s1, Figure S1: Experimental design of the study site. Table S1: The soil physical and chemical properties in 2018 (n = 5). Table S2: The SH and HCI along with soil depth (n = 5). Table S3: Responses of xylem water potential in 2019 (n = 5). Table S4: Results of the correlation analysis between root traits and SH. Table S5: Results of the post-hoc test evaluating the effect of N loading on the relationship between HCI and root proportion under compaction (n = 5). Table S6: The lists of all identified ECM fungal taxon.

Author Contributions: T.S. and T.K. conceived the experiment. T.S., S.Y., Y.T., H.M., and T.K. conducted the experiments, measurements, and data analysis. F.S. managed the experimental site. T.S. wrote the first draft of the paper. Y.T., E.M., T.W., and T.K. jointly revised and edited the paper. All authors have read and agreed to the published version of the manuscript.

Funding: This work was partly funded by the Japan Forest Technology Association, Japan Society for the Promotion of Science, KAKENHI (Proj. No. 18J2013908), and Strategic International Collaborative Research Program of the Japan Science and Technology Agency (Grant No. JPMJSC18HB).

Acknowledgments: We thank K. Ichikawa, E. Fujito and the staff of the Sapporo Experimental Forest of Hokkaido University for the management of nursery. Technical supports from T. Nakaji and E. Agathokleous, and helpful discussion and comments from H. Maruyama, M. Ohashi, T. Yoshida, T. Shinano and the special colleague S. Fujita and R. Doi are gratefully acknowledged. Thanks are also due to HRO Forestry research institute kindly offered seedling samples as a part of their breeding effort as well as E. Paoletti and Y. Hoshika of Institute of Research on Terrestrial Ecosystems of CNR for her kind arrangement of the international cooperation.

Conflicts of Interest: The authors declare no conflict of interest.

References

1. Cambi, M.; Certini, G.; Neri, F.; Marchi, E. The impact of heavy traffic on forest soils: A review. *For. Ecol. Manag.* **2015**, *338*, 124–138. [CrossRef]
2. Marchi, E.; Chung, W.; Visser, R.; Abbas, D.; Nordfjell, T.; Mederski, P.S.; McEwan, A.; Brink, M.; Laschi, A. Sustainable Forest Operations (SFO): A new paradigm in a changing world and climate. *Sci. Total. Environ.* **2018**, *634*, 1385–1397. [CrossRef]
3. Mariotti, B.; Hoshika, Y.; Cambi, M.; Marra, E.; Feng, Z.Z.; Paoletti, E.; Marchi, E. Vehicle-induced compaction of forest soil affects plant morphological and physiological attributes: A meta-analysis. *For. Ecol. Manag.* **2020**, *462*. [CrossRef]
4. Kozlowski, T.T. Soil compaction and growth of woody plants. *Scand. J. For. Res.* **1999**, *14*, 596–619. [CrossRef]
5. Tan, X.; Chang, S.X. Soil compaction and forest litter amendment affect carbon and net nitrogen mineralization in a boreal forest soil. *Soil Tillage Res.* **2007**, *93*, 77–86. [CrossRef]
6. Batey, T. Soil compaction and soil management–a review. *Soil Use Manag.* **2009**, *25*, 335–345. [CrossRef]
7. Kara, Ö.; Bolat, İ. Influence of soil compaction on microfungal community structure in two soil types in Bartin Province, Turkey. *J. Basic. Microbiol.* **2007**, *47*, 394–399. [CrossRef] [PubMed]
8. Moehring, D.M.; Rawls, I.W. Detrimental effect of wet weather logging. *J. For.* **1970**, *68*, 166–167. [CrossRef]
9. Ampoorter, E.; De Frenne, P.; Hermy, M.; Verheyen, K. Effects of soil compaction on growth and survival of tree saplings: A meta-analysis. *Basic Appl. Ecol.* **2011**, *12*, 394–402. [CrossRef]
10. Kormanek, M.; Banach, J.; Leńczuk, D. Influence of soil compaction on the growth of silver fir (*Abies alba* Mill.) under a forest canopy. *Ecol. Quest.* **2015**, *22*, 47–54. [CrossRef]
11. Passioura, J.B. Soil conditions and plant growth. *Plant Cell Environ.* **2002**, *25*, 311–318. [CrossRef] [PubMed]
12. Yoshida, T.; Noguchi, M.; Akibayashi, Y.; Noda, M.; Kadomatsu, M.; Sasa, K. Twenty years of community dynamics in a mixed conifer - broad-leaved forest under a selection system in northern Japan. *Can. J. For. Res.* **2006**, *36*, 1363–1375. [CrossRef]
13. Yamazaki, H.; Yoshida, T. Significance and limitation of scarification treatments on early establishment of *Betula maximowicziana*, a tree species producing buried seeds: Effects of surface soil retention. *J. For. Res.* **2018**, *23*, 166–172. [CrossRef]
14. Fleming, R.L.; Powers, R.F.; Foster, N.W.; Kranabetter, J.M.; Scott, D.A.; Ponder, F., Jr.; Berch, S.; Chapman, W.K.; Kabzems, R.D.; Ludovici, K.H.; et al. Effects of organic matter removal, soil compaction, and vegetation control on 5-year seedling performance: A regional comparison of Long-Term Soil Productivity sites. *Can. J. For. Res.* **2006**, *36*, 529–550. [CrossRef]
15. Ponder, F., Jr.; Fleming, R.L.; Berch, S.; Busse, M.D.; Elioff, J.D.; Hazlett, P.W.; Kabzems, R.D.; Kranabetter, J.M.; Morris, D.M.; Page-Dumroese, D.; et al. Effects of organic matter removal, soil compaction and vegetation control on 10th year biomass and foliar nutrition: LTSP continent-wide comparisons. *For. Ecol. Manag.* **2012**, *278*, 35–54. [CrossRef]
16. Alameda, D.; Villar, R. Moderate soil compaction: Implications on growth and architecture in seedlings of 17 woody plant species. *Soil Tillage Res.* **2009**, *103*, 325–331. [CrossRef]
17. Gebauer, R.; Volarik, D.; Martinkova, M. Impact of soil pressure and compaction on tracheids in Norway spruce seedlings. *New For.* **2011**, *41*, 75–88. [CrossRef]

18. Jourgholami, M.; Khoramizadeh, A.; Zenner, E.K. Effects of soil compaction on seedling morphology, growth, and architecture of chestnut-leaved oak (*Quercus castaneifolia*). *iForest* **2016**, *10*, 145–153. [CrossRef]
19. Cambi, M.; Mariotti, B.; Fabiano, F.; Maltoni, A.; Tani, A.; Foderi, C.; Laschi, A.; Marchi, E. Early response of *Quercus robur* seedlings to soil compaction following germination. *Land Degrad. Dev.* **2018**, *29*, 916–925. [CrossRef]
20. Shah, A.N.; Tanveer, M.; Shahzad, B.; Yang, G.; Fahad, S.; Ali, S.; Bukhari, M.A.; Tung, S.A.; Hafeez, A.; Souliyanonh, B. Soil compaction effects on soil health and crop productivity. *Environ. Sci. Pollut. Res.* **2017**, *24*, 10056–10067. [CrossRef]
21. Reisinger, T.W.; Pope, P.E.; Hammond, S.C. Natural recovery of compacted soils in an upland hardwood forest in Indiana. *North. J. Appl. For.* **1992**, *9*, 138–141. [CrossRef]
22. Bardgett, R. Organism interactions and soil processes. In *The Biology of Soil: A Community and Ecosystem Approach*; Crawley, M.J., Little, C., Southwood, T.R.E., Ulfstrand, S., Eds.; Oxford University Press: Oxford, UK, 2005; pp. 57–85.
23. Zenner, E.K.; Fauskee, J.T.; Berger, A.L.; Puettmann, K.I. Impacts of skidding traffic intensity on soil disturbance, soil recovery, and aspen regeneration in north central Minnesota. *North. J. Appl. For.* **2007**, *24*, 177–183. [CrossRef]
24. Meyer, C.; Lüscher, P.; Schulin, R. Recovery of forest soil from compaction in skid tracks planted with black alder (*Alnus glutinosa* (L.) Gaertn.). *Soil. Tillage Res.* **2014**, *143*, 7–16. [CrossRef]
25. Fernandez, J.L.F.; Hartmann, P.; von Wilpert, K. Planting of alder trees at the edge of skid trails helps to stabilize forest topsoil structure against damage caused by heavy forestry machines. *Soil Tillage Res.* **2019**, *187*, 214–218. [CrossRef]
26. Rodríguez-González, P.M.; Campelo, F.; Albuquerque, A.; Rivaes, R.; Ferreira, T.; Pereira, J.S. Sensitivity of black alder (*Alnus glutinosa* [L.] Gaertn.) growth to hydrological changes in wetland forests at the rear edge of the species distribution. *Plant Ecol.* **2014**, *215*, 233–245. [CrossRef]
27. Schwarz, M.; Lehmann, P.; Or, D. Quantifying lateral root reinforcement in steep slopes - from a bundle of roots to tree stands. *Earth Surf. Process. Landf.* **2010**, *35*, 354–367. [CrossRef]
28. Sinnett, D.; Morgan, G.; Williams, M.; Hutchings, T.R. Soil penetration resistance and tree root development. *Soil Use Manag.* **2008**, *24*, 273–280. [CrossRef]
29. Correa, J.; Postma, J.A.; Watt, M.; Wojciechowski, T. Soil compaction and the architectural plasticity of root systems. *J. Exp. Bot.* **2019**, *70*, 6019–6034. [CrossRef]
30. Aust, W.M.; Burger, J.A.; McKee, W.H., Jr.; Scheerer, G.A.; Tippett, M.D. Bedding and fertilization ameliorate effects of designated wet-weather skid trails after four years for loblolly pine (*Pinus taeda*) plantations. *South. J. Appl. For.* **1998**, *22*, 222–226. [CrossRef]
31. Simcock, R.C.; Parfitt, R.L.; Skinner, M.F.; Dando, J.; Graham, J.D. The effects of soil compaction and fertilizer application on the establishment and growth of *Pinus radiata*. *Can. J. For. Res.* **2006**, *36*, 1077–1086. [CrossRef]
32. Jordan, D.; Ponder, F.; Hubbard, V.C. Effects of soil compaction, forest leaf litter and nitrogen fertilizer on two oak species and microbial activity. *Appl. Soil. Ecol.* **2003**, *23*, 33–41. [CrossRef]
33. Picchio, R.; Tavankar, F.; Nikooy, M.; Pignatti, G.; Venanzi, R.; Lo Monaco, A. Morphology, growth and architecture responses of beech (*Fagus orientalis* Lipsky) and maple tree (*Acer velutinum* Boiss.) seedlings to soil compaction stress caused by mechanized logging operations. *Forests* **2019**, *10*, 771. [CrossRef]
34. Ryu, K.; Watanabe, M.; Shibata, H.; Takagi, K.; Nomura, M.; Koike, T. Ecophysiological responses of the larch species in northern Japan to environmental changes as a basis for afforestation. *Landsc. Ecol. Eng.* **2009**, *5*, 99–106. [CrossRef]
35. Abaimov, A. Geographical distribution and genetics of Siberian larch species. In *Permafrost Ecosystems*; Osawa, A., Zyryanova, O., Matsuura, Y., Kajimoto, T., Wein, R., Eds.; Springer Ecol. Studies 209: Dordrecht, The Netherlands, 2010; pp. 41–58.
36. Kita, K.; Fujimoto, T.; Uchiyama, K.; Kuromaru, M.; Akutsu, H. Estimated amount of carbon accumulation of hybrid larch in three 31-year-old progeny test plantations. *J. Wood. Sci.* **2009**, *55*, 425–434. [CrossRef]
37. Ministry of Agriculture, Forestry and Fisheries (2017) Annual Report on Forest and Forestry in Japan, Fiscal Year 2017 (Summary). Available online: https://www.maff.go.jp/e/data/publish/attach/pdf/index-95.pdf (accessed on 22 June 2020).
38. Reisinger, T.W.; Simmons, G.L.; Pope, P.E. The impact of timber harvesting on soil properties and seedling growth in the south. *South. J. Appl. For.* **1988**, *12*, 58–67. [CrossRef]

39. Cambi, M.; Hoshika, Y.; Mariotti, B.; Paoletti, E.; Picchio, R.; Venanzi, R.; Marchi, E. Compaction by a forest machine affects soil quality and *Quercus robur* L. seedling performance in an experimental field. *For. Ecol. Manag.* **2017**, *384*, 406–414. [CrossRef]
40. Tedersoo, L.; Bahram, M. Mycorrhizal types differ in ecophysiology and alter plant nutrition and soil processes. *Biol. Rev.* **2019**, *94*, 1857–1880. [CrossRef] [PubMed]
41. Wang, X.; Agathokleous, E.; Qu, L.; Fujita, S.; Watanabe, M.; Tamai, Y.; Mao, Q.; Koyama, A.; Koike, T. Effects of simulated nitrogen deposition on ectomycorrhizae community structure in hybrid larch and its parents grown in volcanic ash soil: The role of phosphorous. *Sci. Total Environ.* **2018**, *618*, 905–915. [CrossRef]
42. Sugai, T.; Watanabe, T.; Kita, K.; Koike, T. Nitrogen loading increases the ozone sensitivity of larch seedlings with higher sensitivity to nitrogen loading. *Sci. Total Environ.* **2019**, *663*, 587–595. [CrossRef]
43. Fujita, S.; Wang, X.N.; Kita, K.; Koike, T. Effects of nitrogen loading under low and high phosphorus conditions on above- and below-ground growth of hybrid larch F-1 seedlings. *iForest* **2018**, *11*, 32. [CrossRef]
44. Binkley, D.; Fisher, R.F. Soil management—Harvesting, site preparation, conversion, and drainage. In *Ecology and Management of Forest Soils*, 5th ed.; Binkley, D., Fisher, R.F., Eds.; Temple-Inland Inc.: Hoboken, NJ, USA, 2013; pp. 213–234.
45. WRB, I.W.G. World reference base for soil resources. In *World Soil Resources Report No. 103*; FAO: Rome, Italy, 2006.
46. Research committee of Japanese Institute of Landscape Architecture. Research Committee of Japanese Institute of Landscape Architecture: Ground Maintenance Manual in Landscape Planting. *J. Jpn. Inst. Landsc. Archit.* **2000**, *63*, 224–241.
47. Ministry of the Environment (2019) Annual Report on the Environment in Japan 2019. Available online: https://www.env.go.jp/en/wpaper/2019/index.html (accessed on 22 June 2020).
48. Yamaguchi, M.; Otani, Y.; Li, P.; Nagao, H.; Lenggoro, I.W.; Ishida, A.; Yazaki, K.; Noguchi, K.; Nakaba, S.; Yamane, K.; et al. Effects of long-term exposure to ammonium sulfate particles on growth and gas exchange rates of *Fagus crenata, Castanopsis sieboldii, Larix kaempferi* and *Cryptomeria japonica* seedlings. *Atmos. Environ.* **2014**, *97*, 493–500. [CrossRef]
49. R. Core. Team: A language and environment for statistical computing. R Found. *Stat. Comput. Vienna Austria.* Available online: https://www.r-project.org (accessed on 2 August 2020).
50. Haynes, R.J.; Naidu, R. Influence of lime, fertilizer and manure applications on soil organic matter content and soil physical conditions: A review. *Nutr. Cycl. Agroecosyst.* **1998**, *51*, 123–137. [CrossRef]
51. Gregory, A.S.; Watts, C.W.; Whalley, W.R.; Kuan, H.L.; Griffths, B.S.; Hallett, P.D.; Whitmore, A.P. Physical resilience of soil to field compaction and the interactions with plant growth and microbial community structure. *Eur. J. Soil. Sci.* **2007**, *58*, 1221–1232. [CrossRef]
52. Ampoorter, E.; De Schrijver, A.; De Frenne, P.; Hermy, M.; Verheyen, K. Experimental assessment of ecological restoration options for compacted forest soils. *Ecol. Eng.* **2011**, *37*, 1734–1746. [CrossRef]
53. Fernández, J.F.; Rubin, L.; Hartmann, P.; Puhlmann, H.; von Wilpert, K. Initial recovery of soil structure of a compacted forest soil can be enhanced by technical treatments and planting. *For. Ecol. Manag.* **2019**, *431*, 54–62. [CrossRef]
54. Lambers, H.; Oliveira, R.S. Growth and allocation. In *Plant Physiological Ecology*; Lambers, H., Oliveira, R.S., Eds.; Springer: Berlin/Heidelberg, Germany, 2019; pp. 385–449.
55. Comas, L.H.; Callahan, H.S.; Midford, P.E. Patterns in root traits of woody species hosting arbuscular and ectomycorrhizas: Implications for the evolution of belowground strategies. *Ecol. Evol.* **2014**, *4*, 2979–2990. [CrossRef]
56. Rewald, B.; Rechenmacher, A.; Godbold, D.L. It's complicated: Intraroot system variability of respiration and morphological traits in four deciduous tree species. *Plant Physiol.* **2014**, *166*, 736–745. [CrossRef]
57. Barbier, F.F.; Dun, E.A.; Kerr, S.C.; Chabikwa, T.G.; Beveridge, C.A. An Update on the Signals Controlling Shoot Branching. *Trends Plant Sci.* **2019**, *24*, 220–236. [CrossRef]
58. Tracy, S.R.; Black, C.R.; Roberts, J.A.; Mooney, S.J. Soil compaction: A review of past and present techniques for investigating effects on root growth. *J. Sci. Food. Agric.* **2011**, *91*, 1528–1537. [CrossRef]

Article

Plasticity of Root Traits under Competition for a Nutrient-Rich Patch Depends on Tree Species and Possesses a Large Congruency between Intra- and Interspecific Situations

Zana A. Lak [1,2], Hans Sandén [1], Mathias Mayer [1,3], Douglas L. Godbold [1] and Boris Rewald [1,*]

[1] Department Forest and Soil Sciences, Institute of Forest Ecology, University of Natural Resources and Life Sciences Vienna, 1190 Vienna, Austria; zana.lak@boku.ac.at (Z.A.L.); hans.sanden@boku.ac.at (H.S.); mathias.mayer@boku.ac.at (M.M.); douglas.godbold@boku.ac.at (D.L.G.)

[2] College of Agriculture, Forestry Department, Salahaddin University Erbil, Erbil 44001, Kurdistan Region, Iraq

[3] Forest Soils and Biogeochemistry, Swiss Federal Institute for Forest, Snow and Landscape Research (WSL), 8903 Birmensdorf, Switzerland

* Correspondence: boris.rewald@boku.ac.at; Tel.: +43-1-47654-91219

Received: 1 April 2020; Accepted: 7 May 2020; Published: 9 May 2020

Abstract: Belowground competition is an important structuring force in terrestrial plant communities. Uncertainties remain about the plasticity of functional root traits under competition, especially comparing interspecific vs. intraspecific situations. This study addresses the plasticity of fine root traits of competing *Acer pseudoplatanus* L. and *Fagus sylvatica* L. seedlings in nutrient-rich soil patches. Seedlings' roots were grown in a competition chamber experiment in which root growth (biomass), morphological and architectural fine roots traits, and potential activities of four extracellular enzymes were analyzed. Competition chambers with one, two conspecific, or two allospecific roots were established, and fertilized to create a nutrient 'hotspot'. Interspecific competition significantly reduced fine root growth in *Fagus* only, while intraspecific competition had no significant effect on the fine root biomass of either species. Competition reduced root nitrogen concentration and specific root respiration of both species. Potential extracellular enzymatic activities of β-glucosidase (BG) and *N*-acetyl-glucosaminidase (NAG) were lower in ectomycorrhizal *Fagus* roots competing with *Acer*. *Acer* fine roots had greater diameter and tip densities under intraspecific competition. *Fagus* root traits were generally more plastic than those of *Acer*, but no differences in trait plasticity were found between competitive situations. Compared to *Acer*, *Fagus* roots possessed a greater plasticity of all studied traits but coarse root biomass. However, this high plasticity did not result in directed trait value changes under interspecific competition, but *Fagus* roots grew less and realized lower N concentrations in comparison to competing *Acer* roots. The plasticity of root traits of both species was thus found to be highly species- but not competitor-specific. By showing that both con- and allospecific roots had similar effects on target root growth and most trait values, our data sheds light on the paradigm that the intensity of intraspecific competition is greater than those of interspecific competition belowground.

Keywords: *Acer pseudoplatanus*; competition below ground; extracellular enzymes; *Fagus sylvatica*; intraspecific and interspecific competition; toot economic spectrum; toot respiration; tree root traits

---

1. Introduction

There is strong evidence that belowground competition is an important structuring force in terrestrial plant communities including forests [1–4]—and often at least equally intense as aboveground competition [5–7]. Resource competition below ground is either based on exploitation (i.e., reduction of water and nutrient availability) or interference (i.e., the release of allelopathic compounds inhibiting growth or uptake), governed via fine roots of neighboring trees and their symbionts. In addition, intransitive mechanisms (i.e., competition mediated by factors other than resources, e.g., soil biota) are increasingly considered [8–11]. Competition occurs if negative effects on performance or fitness of a target plant (or components of it) result from the presence of another plant [12]. The magnitude of competitive effects (i.e., competition intensity) is a function of the plant's competitive ability (see [13] for a discussion on competitive response and effect components). Intraspecific competition is often assumed to be more intense due to a nearly complete niche overlap [14]. However, over-proportional reductions of fine root biomass in mixed stands and even over-proliferation of fine roots in soil volumes shared by allospecific roots have been reported [15–17]—pointing towards either greater or reduced root competition intensities under interspecific competition.

Evidence for global patterns of functional variation in plants, such as wood and leaf economic spectra [18,19], indicate that functional traits can enhance our mechanistic understanding of species' trade-offs (i.e., resource acquisition vs. conservation), community composition (e.g., influenced by competitive abilities) and ecosystem function [20,21]. As effects of shoot traits on individual plant strategies are increasingly understood, mechanisms behind species coexistence are progressively unraveled. For example, three key functional traits—wood density, specific leaf area and maximum height—have been shown to consistently influence competitive interactions in trees [22], but are also altered by co-occurring species (e.g., leaf trait variation in beech, [23]). In contrast, variation among functional traits of (tree) fine root systems remains poorly quantified [21,24–26]. Although it becomes increasingly clear that soil resource availabilities are among the key drivers of root trait variation [27,28], few studies have addressed the effects of direct competitive interactions on the plasticity of root functional traits, particularly physiological root traits (e.g., [11,29,30]). This is unfortunate, as trait dissimilarities—as well as increasingly plastic traits and intraspecific trait variations per se—are widely considered to reduce competition intensities and thus increase the probability of species' coexistence [31–34].

As it remains unclear which root traits are most strongly associated with resource limitation [24]—and perhaps also methodological difficulties to quantify resource limitations in (temperate) soils—previous research in forests has largely focused on biomass allocation patterns under competition. Either reduced, increased or unaffected fine root biomasses, depending on tree species' competitive abilities, have been reported [2,17,35]. It remains controversial whether a high degree of trait organization, i.e., one 'root economic spectrum', exists among (tree) fine roots [28,36,37]. However, the suggested traits (e.g., specific root length (SRL), nitrogen (N) content, specific uptake and respiration rates, and lifespan) are strongly related to resource acquisition or conservation [26]—and thus also central to access mechanisms of exploitative competition. In particular, the fast proliferation of active root surface area into heterogeneously distributed, resource-rich patches might be an important strategy to pre-empt resources ([8,38,39], but see [40]). Once space has been occupied, traits related to resource mobilization (such as extracellular enzymes exuded by roots or their fungal symbionts [41]) have to be considered. In sum, more detailed information on the plasticity of root functional traits under competition will facilitate our understanding of plant communities and ecosystem properties.

Mixed stands are increasingly perceived as a way to (partially) mitigate economic and ecological consequences of climate change [42,43]. Thus, the identification of factors which influence their regeneration is key [44–46]. In this context, European beech (*Fagus sylvatica*), the dominant tree species of a potential natural vegetation in large parts of the sub-mountainous altitude range of Central Europe [47], is favored. Sycamore maple (*Acer pseudoplatanus*) is a common competitor of beech—largely sharing its ecological spectrum, but possessing faster growth on resource-rich sites [47]. Successful regeneration of *A. pseudoplatanus* in beech-dominated stands of Central Europe has been attributed to the species' intermediate-to-high shade tolerance coupled with fast height growth if more light becomes available [44,45,48]. *Fagus* and *Acer* are, moreover, associated with different mycorrhizal symbionts, which might influence their nutrient acquisition strategies [49,50]. While ectomycorrhizal symbionts of *Fagus* are able to release enzymes to mobilize nutrients [51], arbuscular symbionts of *Acer* are suggested to (largely) lack these enzymatic abilities [52]. Admixing of beech with maple trees has been recently suggested to foster complementarity effects and to reduce competition within European mixed stands [23]. However, knowledge on the belowground interactions of *F. sylvatica* and *Acer* sp. is rare [30,53,54]—although Simon, et al. [55] underlined the high significance of competitive interactions of *Fagus sylvatica* with other vegetation components on its performance under progressing climate change.

This study aims to identify which root traits of *Acer pseudoplatanus* and *Fagus sylvatica* seedlings adapt under intra- and interspecific competition for a nutrient-rich soil patch. For this purpose, seedlings of both species were grown in a microcosm experiment. Microcosms were connected by competition chambers—wherein inserted roots were set foraging for nutrients. Root traits in intra- and interspecific neighborhoods were compared to roots growing without a competitor (i.e., solation). It is hypothesized that resource competition for a nutrient-rich spot (1) affects growth (biomass) and N concentration of *Acer* and *Fagus* fine roots negatively but to species-specific extents if growing with a con- or allospecific root, i.e., indicating different competition intensities. Furthermore, it is hypothesized that intra- and interspecific competition result in (2) differentiated, directed responses of specific root traits and that fine root biomass is not the sole trait affected, and (3) root trait plasticity is highly species-specific.

2. Material and Methods

2.1. Experiment Set-Up

The experiment was conducted in a ventilated polytunnel greenhouse under slightly increased temperature conditions (approximately 5 °C above ambient daily average) and near ambient lighting conditions (approximately 80%–90% of ambient photosynthetically active radiation, midday; data not shown), in Tulln, Austria (48°19′05.0″ N 16°03′58.2″ E). The experiment took place between mid-April and early September 2017. Two common broad-leaved European tree species of economic importance were studied: *Acer pseudoplatanus* L. (A, Sycamore maple) and *Fagus sylvatica* L. (F, European beech). *Acer pseudoplatanus* is strictly associated with arbuscular mycorrhizal (AM) fungi, while *Fagus sylvatica* is associated with ectomycorrhizal (ECM) fungi [56]. Two-year-old bare-rooted seedlings of similar height (0.9–1.1 m), stem diameter and crown characteristics were obtained in early April 2017 (before budburst) from a local nursery (Natlacen GmbH, Pilgersdorf, Austria); seed certification numbers A/31807-01/2013 (*Acer*) and D092021322711 (*Fagus*). Two-year-old seedlings were chosen because the early developmental stage is crucial for seedling establishment, in particular under competition for limited resources [57], and practical aspects; two-year-old seedlings featured sufficiently long roots while being not too big for the deployed microcosm systems [58].

A microcosm system with two attached 'competition chambers' (CC), allowing for root competitive interactions between seedlings at distinct locations, was developed (Figure 1). The CC are analogous to previously utilized *in situ* competition chambers described in Rewald and Leuschner [17]—allowing us to study intra- and inter-specific root competition effects in a highly controlled manner by minimizing parallel 'non-target' competition (e.g., the co-occurrence of intra- and inter-specific competition in studies using a 'shared'-pot design). In brief, the system comprises two large microcosms (7 L soil) interconnected by two small competition chambers (CC; 1 L soil). Initially, all compartments were filled with a nutrient-poor, sand–silty clay substrate (Supplementary Table S1). One tree seedling was planted per microcosm by mid-April 2017, and one 5-cm long, 'average'-branched (comprising two root orders), terminal fine root axis per plant was carefully inserted into each intra- and interspecific CC; thus, intra- and interspecific CCs hold two fine root axes, inserted from opposite sides (Figure 1). In the isolation treatment (i.e., no competition), one opening (i.e., towards the competitor's microcosm) was sealed and only one fine root segment (of the target species) was inserted. See Supplementary Materials S1 for details.

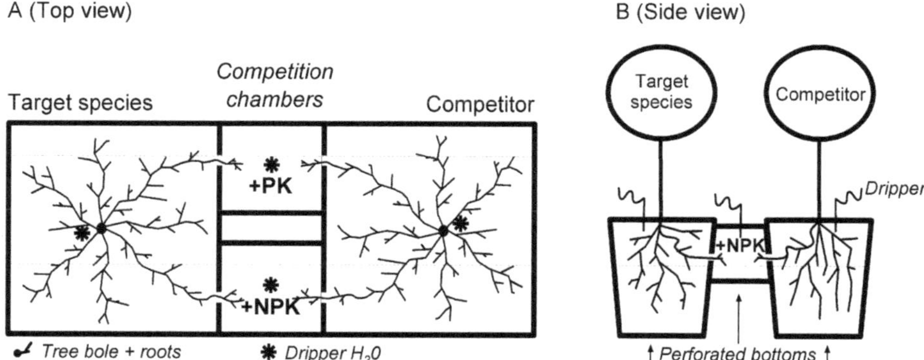

Figure 1. Microcosm system to study competition for localized nutrient-rich soil patches (i.e., competition chambers (CC) with NPK fertilization); (**A**) top view, (**B**) side view. Five-cm long sections of target (focal) species and competitor species' fine roots were inserted from opposite sides into the CC (in intra- and interspecific competition treatments); in the isolation (no competition) treatment, the apertures between the competitor and both CC were closed from one side (not displayed). Positions of tree boles (in microcosms with nutrient-poor substrate) and drippers are displayed. Nutrient-rich fertilizer (+NPK) was applied to the CC only; CC fertilized with a nutrient solution lacking nitrogen (+PK) are not part of this manuscript to keep manageable length and focus. Drawings not at scale.

Thus, three different competition treatments (Table 1) were established per species: (1) isolation (ISO; 'no competition'; only a fine root segment of the target species was inserted into each CC); (2) intraspecific competition (INTRA; roots of two seedlings of the same species, either *Acer* or *Fagus*, were inserted thru opposite sides of the CC); (3) interspecific competition (INTER; roots of one *Fagus* and *Acer* seedling each were inserted from opposite sides of the CC).

Table 1. Experimental set-up with three competitive situations belowground (i.e., isolation, intra- and interspecific competition) for a nutrient-rich spot by roots of two species, *Fagus sylvatica* (F) and *Acer pseudoplatanus* (A), resulting in six treatments (three per species). A:F and F:A originate essentially from the same competition chambers, but a distinction is made whether *Acer* (A:F) or *Fagus* (F:A) is considered the 'target species' (vs. competitor). Abbreviations of treatments are used throughout the manuscript; the number of realized replicates are given.

Type of Root Competition §	Target Species' Root	Belowground Competitor	Treatment (Abbrev.)	Realized Replication (n)
Isolation (ISO; no competition)	*Acer*	none	A	18
	Fagus	none	F	16
Intraspecific competition (INTRA)	*Acer*	*Acer*	A:A	48
	Fagus	*Fagus*	F:F	52
Interspecific competition (INTER)	*Acer*	*Fagus*	A:F	40 *
	Fagus	*Acer*	F:A	39 *

§ Aboveground competition was kept constant with both con- and allospecific neighbors growing at equal distances; * unequal numbers result from one lost *Fagus* sample during lab processing.

An automated, pressure-compensated drip irrigation system was installed ensuring ample water supply. The amount was increased over the growing season in a stepwise manner according to evapotranspiration. The CCs were manually fertilized once per week with 0.05 L of Hoagland solution (+NPK) to create nutrient-rich 'hotspots'. See Supplementary Material S1 for details.

In total, 131 microcosm systems holding 262 trees were set up; 7 were later excluded (see below), resulting in 124 analyzed microcosms (Table 1). A distance of 10 cm was kept between microcosms. The microcosms were set up randomly and in alternating directions (i.e., target species either placed north or south of competitor), to ensure a homogeneous competition environment above ground. After four months of fertilization (May–August), the CCs were harvested in early September 2017.

2.2. *Harvesting of NPK Fertilized Competition Chambers*

During harvest, the roots were (i) cut at the competition chambers' (CC) apertures (from the inside towards the microcosm), (ii) the bottom of the CC was opened and the substrate was emptied into a bowl, and (iii) the root origin (i.e., from one of the two microcosms) was marked. Subsequently, the apertures (towards the microcosms) were investigated for additional root in-growth (i.e., beside the initially inserted root axis), and roots were investigated for viability using morphological criteria and color [17]. If additional root in-growth or dead roots were detected, the CC were excluded from further analyses (7 CC in total), see realized replicate numbers (Table 1). While induced root death might be a competition mechanism [29], it cannot be ruled out that dead roots resulted from damage that occurred during installation. The roots were carefully rinsed under tap water and untangled into separate root branches—according to the microcosm of origin—keeping the roots moist at all times. No spatial segregation of roots in the CC was noticed during harvest. Subsequently, coarse root segments (diameter (d) > 2mm) were manually dissected from fine roots (d ≤ 2 mm) using a caliper and stored separately.

2.3. *Specific Fine Root Respiration*

Within 10 min after harvesting, surface-moist fine root sub-samples were entered into 55 mL plastic tubes with lids and placed in a climate cabinet (20 °C) for temperature acclimation (~10–15 min). Subsequently, the roots were blotted surface dry and the CO_2 efflux was determined at 20 °C with an infra-red gas-analyzer (IRGA; EGM-5, PP-Systems International, Inc. Amesbury, MA, USA) as ΔCO_2 (ppm)—recording ppm values every 30 sec for 4–5 minutes (+60 s 'deadband' to stabilize CO_2 readings before measurements). Slopes of ΔCO_2 were calculated by linear regressions (as no curve flattening, requiring polynomial approaches, was observed); specific (fine) root respiration rates (RR_S; nmol CO_2 g^{-1} s^{-1}) were calculated, taking headspace, air pressure, temperature, and root dry mass (see below) into account.

2.4. *Potential Enzymatic Activity*

For measuring the potential enzymatic activity (PEA) of root tips, 4–5 root tips (~2 mm in length) were randomly sampled per root branch within 10 minutes after harvesting. Root tips were placed in reaction tubes filled with deionized water, and stored for 14–20 h at 4–5 °C. Two of the sampled tips were used the next morning to determine PEAs (nmol cm^2 h^{-1}) at the rhizoplane, using high-throughput photometric and fluorometric microplate assays [59,60]. Four enzymes were measured: acid phosphatase (PEA_{AP}, releasing inorganic phosphate from organic matter), β-glucosidase (PEA_{BG}, hydrolyzing cellobiose into glucose), leucine-amino-peptidase (PEA_{LAP}, breaking down polypeptides), and *N*-acetyl-glucosaminidase (PEA_{NAG}, breaking down chitin). Methylumbelliferone-complexed substrates were used for AP (4-methylumbelliferyl phosphate), BG (4-methulumbelliferyl β-D-glucopyranoside) and NAG (4-methylumbelliferyl *N*-acetyl-β-glucosaminide); 7-amino-4-methylcoumarin (AMC) was used for LAP (L-leucine-7-amido-4-methylcoumarin hydrochloride). All root tips were subsequently imaged and analyzed for surface area (see below).

2.5. Root Morphology, Biomass and Root Competition Intensity

Fine (d ≤ 2 mm) and coarse (d > 2mm) root samples (including samples for respiration measurements) were stored in tap water at 4–5 °C until further processing. Fine root samples were individually imaged with a flatbed scanner (Expression 10000XL with transparency unit, Epson, Japan; 600 dpi, grey-scale). Images were analyzed with the software WinRhizo 2012b Pro (Regent, Quebec, Canada) for morphological root traits including length (cm), surface area (cm^2), volume (cm^3), and average diameter (RD; mm). In addition, root tip density (RTD; n cm^{-1}) was calculated by dividing recorded tip numbers by length.

Root samples were dried to constant mass (65 °C) and weighed (±0.1 mg); fine root (FRB) and coarse root (CRB)) values were recorded (g DM, dry matter). Specific root area (SRA; cm^2 g^{-1}) and tissue density (TD; g cm^{-3}) were calculated for dried fine root samples. In addition, total FRB (i.e., sum of target and competitor fine root biomass) was calculated per CC with intra- and interspecific competition.

To measure the strength of competition, three relative competition intensity (RCI) indices were calculated for each species using the fine root biomass (FRB). $\overline{RCI}_{Intra\ vs\ Iso}$ standardizes fine root biomass (FRB) in intraspecific mixtures (INTRA; i.e., monoculture) with FRB in isolation (ISO, i.e., 'alone', Equation (1); this study), $\overline{RCI}_{Inter\ vs\ Iso}$ standardizes FRB in interspecific mixtures (INTER) with FRB in isolation (Equation (2); sensu [61]), and $\overline{RCI}_{Inter\ vs\ Intra}$ standardizes FRB in interspecific mixtures with FRB in intraspecific mixtures (Equation (3); sensu [62]). Species-specific means ($\overline{FRB}$) under isolation (Equations (1) and (2)) or under intraspecific competition (Equation (3)) were used to calculate mean RCI values by comparison with individual FRB values for n competition chambers (n = 39–52).

$$\overline{RCI}_{Intra\ vs\ Iso} = \frac{1}{n}\sum_{i=1}^{n} (\overline{FRB}_{ISO} - FRB_{INTRA}) \ / \ \overline{FRB}_{ISO} \quad (1)$$

$$\overline{RCI}_{Inter\ vs\ Iso} = \frac{1}{n}\sum_{i=1}^{n} (\overline{FRB}_{ISO} - FRB_{INTER}) \ / \ \overline{FRB}_{ISO} \quad (2)$$

$$\overline{RCI}_{Inter\ vs\ Intra} = \frac{1}{n}\sum_{i=1}^{n} (\overline{FRB}_{INTRA} - FRB_{INTER}) \ / \ \overline{FRB}_{INTRA} \quad (3)$$

2.6. Root and Soil Chemical Analysis

For determination of root carbon (C) and nitrogen (N) concentrations, dried (70 °C, until constant mass) fine roots were ground to powder (Pulverisette 5; Fritsch, Idar-Oberstein, Germany). Fine roots were analyzed by pooling and homogenizing three random samples each per treatment (n = 5). Total C and N concentrations (mg g^{-1}) were determined by dry combustion using a TruSpec CN analyser (Leco, St. Joseph, USA) according to the Austrian ÖNORM L1080 protocol. C:N ratios were calculated. See Supplementary Material S2 for details on soil chemical analysis.

2.7. Root Trait Plasticity

The plasticity of root traits under the different competitive situations was calculated as relative distance plasticity index (RDPI) with strong statistical power to test for differences in plasticity of traits within and between species (see [63] for details). Root phenotypic plasticity was determined for the following (normalized) traits: FRB, CRB ('biomass traits'); RD, TD, SRA and RTD ('morphological traits'); and RR_S, PEA_{AP}, PEA_{BG}, PEA_{LAP}, PEA_{NAG} ('physiological traits'). RDPI(X) values were calculated for each specific trait (X) as relative phenotypic distances between individuals (d_{ij}→i'j') of the same species under different competition treatments, divided by the sum ($x_{i'j'}$ + x_{ij}). An RDPI ranging from 0 (no plasticity) to 1 (maximal plasticity) was obtained for each species as:

$$RDPI = \Sigma(d_{ij} \rightarrow i'j'/(x_{i'j'} + x_{ij}))/n, \quad (4)$$

where n is the total number of distances. Three 'types' of RDPI were calculated: (1) $RDPI_{Total}$, considering relative phenotypic distances between all three competition treatments (i.e., ISO | INTRA | INTER), (2) $RDPI_{INTRA\ vs\ ISO}$, considering relative phenotypic distances between traits under intraspecific competition and isolation only, and (3) $RDPI_{INTER\ vs\ ISO}$ considering relative phenotypic distances between traits under interspecific competition and isolation only. RDPI values were calculated by means of the statistical software R, v. 3.5.3, [64] and the R package 'Plasticity' [65]. Subsequently, $RDPI_{Total}$ values of root traits were compared between species. $RDPI_{INTRA\ vs\ ISO}$ and $RDPI_{INTER\ vs\ ISO}$ values of specific traits were contrasted to determine if the plasticity of all or some traits differs between intra- and interspecific competition.

2.8. Statistical Analysis

Statistical analysis was performed using the PC software SPSS v. 24.0 (SPSS, IL, USA) and Microsoft Excel 2013. The data were transformed to obtain a normal distribution when needed (Kolmogorov–Smirnov test) and controlled for homogeneity of variances (Leven's test). Overall differences in root traits between species and competitive situations were tested with a general linear model (GLM) with species and competition as fixed variables (see Supplementary Material S4 for GLM statistics). In addition, all traits were tested for differences between treatments with a *t*-test; RCI was tested for differences between species within competitive environments with a *t*-test; $RDPI_{Total}$ values among species and/or competition treatments were analyzed by GLMs followed by two-sample *t*-tests. Please note, the alpha-value of *t*-tests were not Bonferroni-corrected, following the arguments of Moran [66]. All data represent mean ± standard error (SE). Statistical relationships were considered significant at $p < 0.05$.

3. Results

3.1. Root Biomass and Competition Intensity

The fine (FRB; Figure 2a; Supplementary Table S2), coarse (CRB; Supplementary Figure S1, Table S3), and total root biomass (Supplementary Table S4) of *Acer* were larger than those of *Fagus*. In *Acer*, FRB did not vary significantly under intra- and interspecific competition compared to isolation (i.e., no competition). In contrast, significantly less fine and coarse root biomass were found in *Fagus* under interspecific competition (F:A) compared to isolation. The relative distance plasticity indices ($RDPI_{Total}$; Figure 3a) of FRB and CRB possessed in both species the greatest values across all measured traits (beside $RDPI_{Total}(PEA_{NAG})$ of *Fagus*, see below; Figure 3a).

$RDPI_{Total}$(FRB) of *Fagus* was significantly greater than that of *Acer*, while $RDPI_{Total}$(CRB) values did not differ between species. The RCI values of *Acer* were generally lower than RCI values of *Fagus*; $RCI_{INTER\ vs\ ISO}$ of *Fagus* under interspecific competition tended ($p = 0.06$) to be greater than $RCI_{INTER\ vs\ ISO}$ of *Acer* (Figure 2b).

3.2. Fine Root Respiration

The specific fine root respiration (RR_S) rates were significantly lower in both species grown under intra- and interspecific competition in comparison with isolation (Figure 4). In most cases, RR_S did not differ significantly between species within the same competition treatment; however, the RR_S of *Acer* under intraspecific competition (A:A) was significantly lower than the RR_S of *Fagus* under intraspecific competition (F:F). Within species, the *t*-test indicated no significant differences in RR_S between intra- and interspecific competition, however, a significant interaction in the GLM between species and competition type may hint at *Acer* fine roots respiring more under interspecific competition and *Fagus* under intraspecific competition (Supplementary Table S5).

$RDPI_{Total}$ (RR_S) values of *Acer* and *Fagus*' RR_S differed significantly between species (Figure 3a).

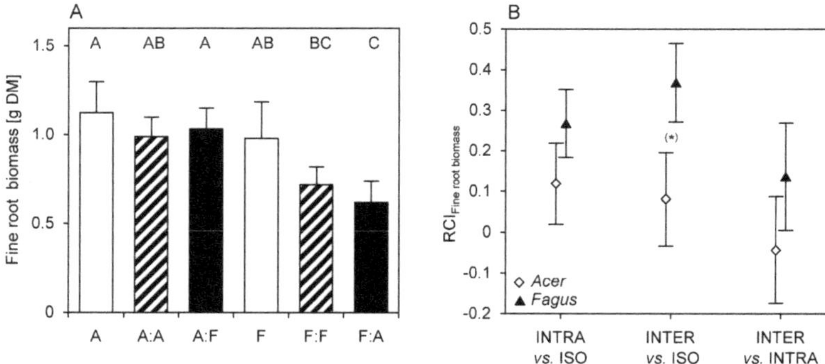

Figure 2. (**A**) Biomass (g DM, dry matter) of fine roots of *Acer pseudoplatanus* (A) and *Fagus sylvatica* (F) grown under three different competition treatments each into nutrient-rich soil patches. A, *Acer* root grown in isolation (no competition); A:A, *Acer* root grown in competition with another *Acer* root; A:F, *Acer* grown in competition with *Fagus*; F, *Fagus* grown in isolation; F:F, *Fagus* grown in competition with *Fagus*; F:A, *Fagus* grown in competition with *Acer*. Significant differences between treatments are indicated by different letters (*t*-test, $p < 0.05$; mean + standard error (SE), n = 39–52; see Supplementary Table S2 for GLM statistics on log-transformed FRB data). (**B**) Relative competition intensity (RCI) based on fine root biomass (FRB) of *Acer* (open rectangles) and *Fagus* (filled triangles) under 1) intraspecific competition (INTRA) relative to isolation (ISO), 2) interspecific competition (INTER) relative to ISO, and 3) INTER relative to INTRA. Tendencies between species are indicated by (*) (*t*-test, p = 0.06; mean ± SE, n = 39–52).

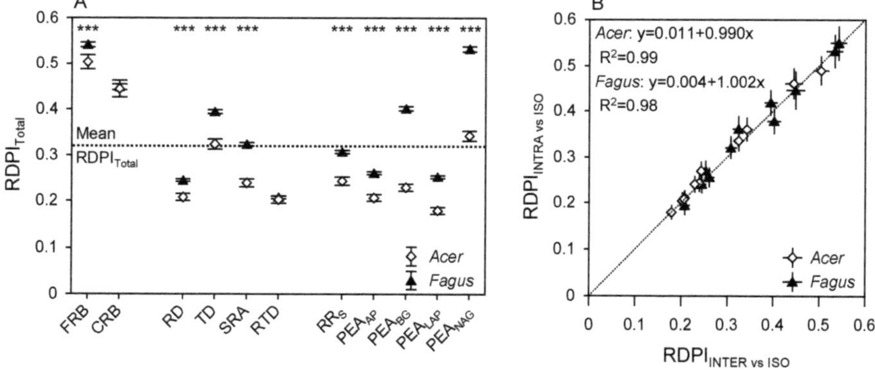

Figure 3. **(A)** Relative distance plasticity indices ($RDPI_{Total}$) for the traits fine root biomass (FRB), coarse root biomass (CRB), root diameter (RD), tissue density (TD), specific root area (SRA), root tip density (RTD), specific root respiration (RR_S) and the potential enzymatic activities (PEA) of acid phosphatase (PEA_{AP}), β-glucosidase (PEA_{BG}), leucine-amino-peptidase (PEA_{LAP}), and N-acetyl-glucosaminidase (PEA_{NAG}) of *Acer pseudoplatanus* (open rectangles) and *Fagus sylvatica* (filled triangles) across isolation, intra- and interspecific competition. The mean $RDPI_{Total}$ value (across traits and species) is shown as the dotted line. Significant differences between species are marked (*t*-test, $p < 0.001$ ***; mean ± three standard errors (3SE)). **(B)** Relative distance plasticity indices under intraspecific competition ($RDPI_{INTRA\ vs\ ISO}$) vs. RDPI under interspecific competition ($RDPI_{INTER\ vs\ ISO}$; mean ± 3SE); linear regressions (formulae and R^2) of trait means are given for *Acer* and *Fagus*; the 1:1 line is drawn for comparison (dotted line).

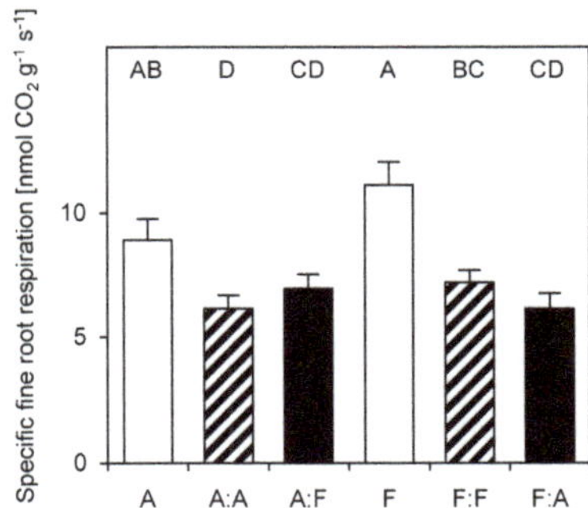

Figure 4. Specific fine root respiration (RR_S; 20 °C) of *Acer pseudoplatanus* (A) and *Fagus sylvatica* (F) under three different competition treatments in nutrient-rich soil patches. A, *Acer* root grown in isolation (no competition); A:A, *Acer* root grown in competition with another *Acer* root; A:F, *Acer* grown in competition with *Fagus*; F, *Fagus* grown in isolation; F:F, *Fagus* grown in competition with *Fagus*; F:A, *Fagus* grown in competition with *Acer* (mean + SE, n = 15–49; see Supplementary Table S5 for GLM statistics).

3.3. Potential Extracellular Enzymatic Activities

Fagus root tips possessed generally higher potential enzymatic activities (PEA) at the surface than *Acer* root tips (Figure 5). Specifically, the PEA of acid phosphatase (PEA_{AP}) possessed no differences to either intra- or interspecific competition compared to isolation (Figure 5a). Similarly, the PEA of leucine-amino-peptidase (PEA_{LAP}) was unaffected by the competition treatments in both species (Figure 5c). Finally, the PEAs of both β-glucosidase (PEA_{BG}) and *N*-acetyl-glucosaminidase (PEA_{NAG}) were unaffected by competition in *Acer* while being significantly reduced in *Fagus* under interspecific competition (F:A) compared to isolation (Figure 5b,d).

$RDPI_{Total}$ values of PEA_{AP} and PEA_{LAP} ranged in both species from 0.18–0.26; the plasticity of PEA_{BG} had a similar extent compared to PEA_{AP} and PEA_{LAP} in *Acer* but was greater in *Fagus* (Figure 3a). PEA_{NAG} possessed $RDPI_{Total}$ values of 0.34 ± 0.00 and 0.53 ± 0.01 in *Acer* and *Fagus*, respectively.

3.4. Fine Root Morphology

The average fine root diameter (RD) differed significantly between *Acer* grown under intra- (A:A) compared to interspecific (A:F) competition, with lower average diameters under interspecific competition (Figure 6a). The tissue density (TD) and specific root area (SRA) of *Acer* fine roots were generally lower or greater, respectively, compared to *Fagus*; competition had no significant influence on either trait (Figure 6b,c). GLM evidenced no overall competition effect on root diameter (RD), tissue density (TD) and specific root area (SRA; Supplementary Tables S6–S8). The root tip density was generally greater under intraspecific competition as compared to isolation and intraspecific competition (Supplementary Table S9); this effect was largest in *Acer* (Figure 6d). *Fagus* had both higher TD and RTD than *Acer* (Supplementary Tables S8 and S9).

The RD and RTD of both species possessed relatively low plasticity indices under competition, with $RDPI_{Total}$ values of 0.20–0.24; however, $RDPI_{Total}$ (RD) differed significantly between species, with greater RDPI values in *Fagus* (Figure 3a). The $RDPI_{Total}$ (TD) differed significantly between *Acer* and *Fagus*. Similar, $RDPI_{Total}$ (SRA) values differed significantly between species, with a significantly greater plasticity of SRA in *Fagus* compared to *Acer*.

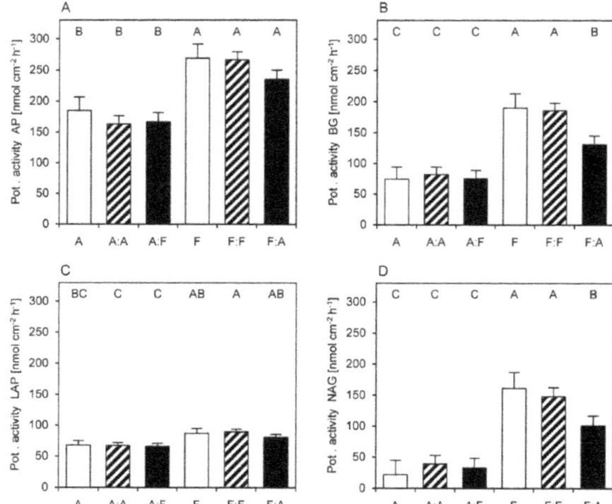

Figure 5. Potential enzymatic activities (PEA; nmol cm^{-2} h^{-1}) of **(A)** acid phosphatase (AP), **(B)** β-glucosidase (BG), **(C)** leucine-amino-peptidase (LAP), and **(D)** *N*-acetyl-glucosaminidase (NAG) on the rhizoplane of *Acer pseudoplatanus* (A) and *Fagus sylvatica* (F) root tips growing under three different competition treatments each in nutrient-rich soil patches. A, *Acer* root grown in isolation (no competition); A:A, *Acer* root grown in competition with another *Acer* root; A:F, *Acer* grown in competition with *Fagus*; F, *Fagus* grown in isolation; F:F, *Fagus* grown in competition with *Fagus*; F:A, *Fagus* grown in competition with *Acer*. Significant differences between treatments are indicated by different letters (*t*-test, $p < 0.05$, mean + SE, n = 14–50).

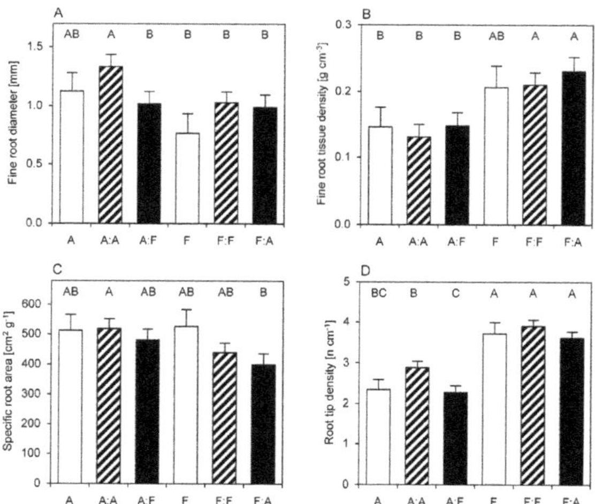

Figure 6. Morphological traits of fine roots. **(A)** Average fine root diameter (RD), **(B)** fine root tissue density (TD), **(C)** specific root area (SRA), and **(D)** root tip density (RTD) of *Acer pseudoplatanus* (A) and *Fagus sylvatica* (F) grown under three different competition treatments into nutrient-rich soil patches. A, *Acer* root grown in isolation (no competition); A:A, *Acer* root grown in competition with another *Acer* root; A:F, *Acer* grown in competition with *Fagus*; F, *Fagus* grown in isolation; F:F, *Fagus* grown in competition with *Fagus*; F:A, *Fagus* grown in competition with *Acer*. Significant differences between treatments are indicated by different letters (*t*-test, $p < 0.05$, mean + SE, n = 16–52; see Supplementary Tables S6–S9 for GLM statistics, partially on log-transformed data).

3.5. Fine Root N Concentrations

No significant differences in total nitrogen (N) concentrations were found between *Acer* and *Fagus* fine roots grown under similar competitive treatments. However, N concentrations were significantly reduced by approximately 7%–33% under competition in both species (Figure 7, Supplementary Table S10). In *Fagus*, the N under interspecific competition (F:A) was significantly lower compared to both intraspecific competition (F:F) and isolation (F), while in *Acer*, the N concentration under interspecific competition was significantly lower compared to isolation only. The low N content under interspecific competition is reflected by a greater C:N ratio under interspecific competition compared to the other two treatments; see Supporting Material S2 for fine root total carbon concentration and C:N ratios (Supplementary Figure S2, Table S11).

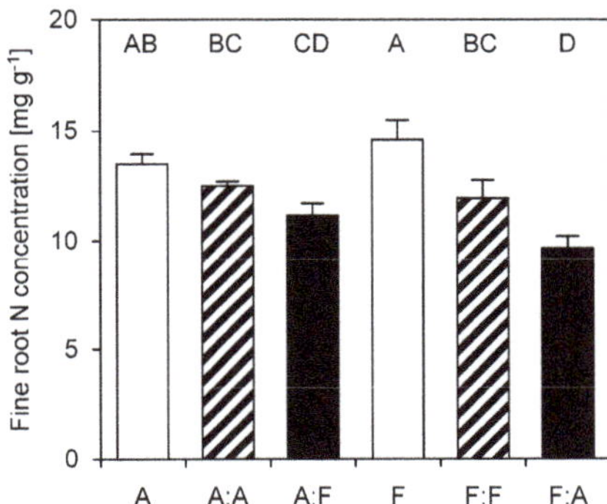

Figure 7. Fine root nitrogen (N) concentration of *Acer pseudoplatanus* (A) and *Fagus sylvatica* (F) grown under three different competition treatments into nutrient-rich soil patches. A, *Acer* root grown in isolation (no competition); A:A, *Acer* root grown in competition with another *Acer* root; A:F, *Acer* grown in competition with *Fagus*; F, *Fagus* grown in isolation; F:F, *Fagus* grown in competition with *Fagus*; F:A, *Fagus* grown in competition with *Acer*. Significant differences between treatments are indicated by different letters (*t*-test, $p < 0.05$, mean + SE, n= 4–5; see Supplementary Table S10 for GLM statistics).

3.6. Plasticity Index

Greater relative distance plasticity index (RDPI) values indicate a greater plasticity of a specific trait under different environmental conditions, i.e., here understood as the different competition treatments. *Fagus* roots possessed greater $RDPI_{Total}$ values for the traits FRB, RD, TD, SRA, RR_S and all four PEAs than *Acer* (Figure 3a). For both species, the $RDPI_{Total}$ values for the traits FRB and CRB were above the mean of all traits' $RDPI_{Total}$ across species (0.32 ± 0.02; dotted line in Figure 6b); $RDPI_{Total}$ values of TD, PEA_{BG} and PEA_{NAG} were 30%–50% greater than the mean $RDPI_{Total}$ in *Fagus* only. In both species, the traits RD, RTD, PEA_{AP} and PEA_{LAP} possessed $RDPI_{Total}$ values well below the mean.

The $RDPI_{INTRA\ vs\ ISO}$ and $RDPI_{INTER\ vs\ ISO}$ values per species and trait were comparable to the respective $RDPI_{Total}$ values, but slightly lower (Figure 3). The RDPI values under intra- and interspecific competition, each relative to the trait values under isolation (Control), were highly related and thus aligned closely to a hypothesized 1:1 line (Figure 3b).

4. Discussion

4.1. Influence of Competition for a Nutrient-Rich Soil Spot on Fine Root Foraging Behaviour, Root Nitrogen Status and Root Trait Characteristics

The root biomass production rates, root N concentrations and C:N ratios measured in the competition chambers (CC) were comparable to values from nutrient-rich top soil layers in mature stands dominated by either species [67], suggesting that realistic experimental conditions were established. No signs of fine root over-proliferation were found [38] but individual root biomasses were in general lower in CCs with a competing root compared to isolation (Figure 2a). The stronger decrease in fine root biomass in *Fagus* as compared to *Acer* seedlings under competition and the relative competition intensity (RCI) indices illustrate that *Fagus* seedlings' roots are affected to a greater extent by roots sharing the same soil volume than *Acers'* (Figure 2b). In accordance, significant effects of neighboring plants on *Fagus* root systems (i.e., reduced total fine root biomass/root length density or shifted rooting depths) were previously reported for e.g., mature mixed stands of *Fagus* and *Picea abies* [35,68,69], although with contrasting results in regard to the shift of *Fagus'* fine root production to deeper or more shallow soil horizons, respectively. In contrast, Leuschner, Hertel, Coners and Büttner [2] reported increasing *Fagus* fine root biomass (i.e., over-proliferation) when competing with *Quercus petraea* roots for N-rich top soil layers, concluding that this competitive replacement of *Quercus* fine roots by faster growing *Fagus* roots indicates asymmetric interspecific root competition in favor of *Fagus* in the studied stand. Although data on *Acer* sp. fine root biomass in monocultures is absent, to the best of our knowledge, Meinen, et al. [70] showed that the (relative) fine root biomass of mature *Acer* sp. increased with increasing tree species diversity level *in situ*. Based on findings on mature trees, we had thus hypothesized that competition intensities between seedling roots also differ largely between competitive situations. The supposed difference between intra- and interspecific situations is usually related to two contrasting ideas—one being that conspecific roots compete for more similar resources and root growth is thus inhibited to a larger extent compared to interspecific situations, where facilitative aspects may dominate [17,71], and another being that plants may recognize their 'kin' and compete less with conspecifics vs. 'strangers', i.e., allospecific roots [32,72]. However, our data does not support either hypotheses for both species as fine root biomasses and competition intensities under intra- and interspecific competition did not differ significantly, indicating similar effects of neighboring roots independent of their identity. However, as the fine root biomass of *Fagus* declined significantly compared to isolation while *Acers'* did not, the available data underscore that root foraging behaviors under interspecific competition are in favor of *Acer* seedlings. Together with previous results, using similar experimental set-ups in situ [17,71,73], our results add to the conclusion that root foraging behaviors (of temperate trees) in shared soil patches are highly species-specific and modulated by the respective environmental conditions [30,71,74]. We thus suggest that our current understanding of root competitive interactions does not yet allow for drawing general predictions on the intensity of root interactions under intra- vs. interspecific competition between species.

While a reduced root resource uptake capacity, as related to reduced fine root biomasses, length or surface, might be 'counterbalanced' in theory by increasing specific uptake rates, the fine root N concentrations of both *Fagus* and *Acer* seedlings decreased under interspecific competition while being significantly lower under intraspecific competition in *Fagus* (Figure 7). Similarly, Simon, et al. [75] reported that under interspecific competition, the inorganic N uptake rates (per root dry weight) decreased by up to 80% in *Fagus* but increased by 30%–50% in *Acer* seedlings, resulting in significantly lower N concentrations in the roots of *Fagus* compared to intraspecific situations. The results of Simon and colleagues are also consistent with earlier reports stating 30%–60% lower/decreased inorganic N accumulation efficiencies/uptake rates in *Fagus* seedlings under competition with *Rubus fruticosus* [74,76]. While we cannot exclude modified N translocation rates (to the shoot or other parts of the root system), our results on root N concentrations are in line with previous reports indicating lower N uptake capacities of *Fagus* roots in comparison to *Acer* seedlings' roots.

Specific root respiration (RR_S) rates depend on three major energy-requiring processes, namely ion uptake and mobilization, growth and defense, and maintenance of living cells, and root respiration represents a major sink of assimilated C [77]. As both species possess relatively similar growth rates at sapling stage [44,48], and RR_S is generally considered to be related to growth rates [77], the similarity of RR_S of *Fagus* and *Acer* within the same competitive situation might come as a limited surprise. However, our results are in contrast to findings on *Pisum sativum* and root tips of *Larix gmelinii*, where (nocturnal) root respiration increased significantly under non-self/interspecific competition for unknown reasons (see [11,78] and the discussion within). Furthermore, our experiment did not find significant differences in RR_S between intra- and interspecific competition (Figure 4), while Zwetsloot, Goebel, Paya, Grams and Bauerle [69] recently reported that oxygen consumption rates of *Fagus*, and partially *Picea abies*, fine roots (of mature trees during spring) were significantly lower under interspecific competition compared to 'single species' conditions. However, it remains open if Zwetsloot and colleagues measured 'single species' RR_S on roots competing with roots of conspecifics, the same individual, or isolated roots. As respiration is highly related to root N concentrations—as a proxy for the amount of protein—it may serve as a predictor of root tissue activity [26]. Thus, lower root N concentrations and the sum of reduced, RR_S-effective 'activities' such as growth and exudation, may underlie the reduced RR_S rates under competition found in our study. Further studies are needed to untangle the contrastingly reported, potentially direct (e.g., interference competition) or indirect (e.g., resource competition) effects of roots sharing the same soil volume on specific root respiration rates.

In addition to root biomass and root respiration, root exudates (directly or via C transfer to exuding mycorrhiza) can be a substantial sink for assimilated carbon and have a major influence on plant mineralization and nutrient uptake capacity and efficiency. The C investment in symbiotic microorganisms, in our case ectomycorrhiza for *Fagus* and arbuscular mycorrhiza for *Acer*, is reflected in the potential enzymatic activity (PEA) on the root rhizoplane—as among the four analyzed PEAs, roots can only produce phosphatase (AP, Figure 5a). In contrast, NAG in particular has been found to be strongly related to soil fungal biomass [see [79] and references within). The generally higher PEAs in *Fagus* are likely based on the higher enzyme exudation rates of ECM compared to AM fungi [80]; NAG activity has been found to be strongly correlated to Basidiomycota and Ascomycota (both ectomycorrhizal phyla). Changes in PEAs of BG, LAP and NAG are thus likely (co-)related to changes in mycorrhizal colonization rates or identity of the symbionts; both parameters were previously reported to differ between monocultures and mixtures [11,81], but lay beyond the focus of this study. A lower colonization rate of *Fagus* under interspecific competition could thus be another factor underlying the lower extracellular enzymatic activity (reduced PEA_{NAG} and PEA_{BG}) of *Fagus* fine roots and their symbionts under interspecific competition. In contrast, an increase in the enzymatic activity, especially of NAG, would have indicated a change in nutrient foraging strategy from roots to hyphae under competition—however, our study does not provide evidence for such potentially adaptive changes in neither of the two species.

In sum, we could not generally confirm our first hypothesis, namely that that sharing a nutrient-rich spot with another root strongly affects root biomass and N status of fine roots negatively; *Acers'* root growth and N content were, especially under intraspecific competition, affected only to a minor extent. However, our data does provide support for the hypotheses that root competition intensities differ for the two studied tree species but do not confirm that competitive interactions generally differ under intra- and interspecific situations. This strongly contrasts common findings above ground, namely that 'competition within species [is] stronger than between species' [22]—the divergence is likely related to the multitude of resources below- compared to aboveground.

4.2. *Species-Specific Plasticity of Functional Root Traits (under Intra- and Interspecific Competition)*

In our third hypothesis, we speculated that different competitive neighborhoods trigger distinct, species-specific responses among root traits—potentially increasing the differences between specific root traits. Our results evidence that root traits of *Fagus* seedlings were in general significantly more plastic than *Acer* root traits (Figure 3). This fits with previous studies, which frequently describe the root system (biomass) of *Fagus sylvatica* as being very dynamic and adaptable to competitive situations compared to other Central European tree species (e.g., [17,68,69]). Specifically, we found a high degree of plasticity in biomass-related traits (i.e., FRB, CRB) of both studied species while the morphological root traits studied in this work (i.e., RD, TD, SRA), especially of *Fagus*, possessed limited plasticities. While the high plasticity of fine root biomass fits previous findings on tree roots competing with other tree roots [17], it contrasts the findings of e.g., *Fagus* with herbs where FRB was not responsive [82].

The low plasticity of morphological traits was surprising as it has been frequently hypothesized that e.g., the production of thinner roots with a greater SRL or SRA in response to a specific neighbor could improve nutrient and water uptake under competition [83,84]. Indeed, increased SRL were previously reported for competing, mature *Fagus* trees [68], and *Fagus* seedlings competing with herbs [82]. Furthermore, consistent responses to N enrichment resulting in greater fine root diameters of temperate trees has been reported recently [85]. However, similar to our findings (Figure 6a,c), Lei et al. [86] reported no significant differences in *Fagus* fine-root diameter and SRL between different species richness levels. In *Acer*, on the other hand, fine root diameter (RD) and root tip density (RTD) were significantly greater under intra- compared to interspecific competition (Figure 6a,d). As we did not perform a root-order based analysis, we can only speculate that the increased mean fine RD in *Acer* is related to the increase in RTD, as 'swollen' root tips often feature a slightly larger diameter than the next higher fine root orders. As the N content of *Acer* fine roots was sustained under intraspecific competition, this might point to a benefit of a greater density of 'physiological active' root tips for N uptake [87]; however, studying the underlying (e.g., anatomical) traits of the RD change in greater detail would be necessary to draw general conclusions [85]. Among the physiological root traits studied, *Fagus* showed a high plasticity in PEA_{NAG}. As PEA_{NAG} is related to the presence of fungal symbionts, and ectomycorrhiza are only present in *Fagus*, the high plasticity probably largely reflects different ectomycorrhizal colonization or activity as discussed above. In contrast, $RDPI_{Total}$ values do indicate a generally low plasticity of RR_S and other PEAs. No consistent patterns regarding the influence of competition for local nutrient patches on morphological or physiological root traits have thus emerged in our study.

In summary, our study possesses limited evidence for greater root trait dissimilarities under interspecific competition. This is further supported by comparing the plasticities between isolation and intra- ($RDPI_{INTRA\ vs\ ISO}$), and isolation and interspecific ($RDPI_{INTER\ vs\ ISO}$) competition. They follow a near 1:1 pattern in our study—i.e., indicating very similar trait plasticities irrespective of the competing species (Figure 3b). Thus, most observed changes in root trait values under competition compared to isolation might be rather a 'passive' reaction to resource availability (or a modified stoichiometry) and not directed or even adaptive (to specific competitors)—in the sense of increasing the competitive ability of a specific root. However, as information on root trait values is still scarce in general and not related to different competitive situations but environmental gradients at best [27,28], it seems too early to draw general conclusions on root reactions norms under resource competition. However, due to the marked differences in root traits under isolation compared to 'competition', we suggest that information on the competitive neighborhood (in a shared soil space) is key ancillary data needed to better interpret functional root traits deposited within databases [88,89].

5. Conclusions

In accordance with previous results [30,53,75], our study underscores the inability of roots of *Fagus sylvatica* seedlings to successfully compete with *Acer pseudoplatanus* roots for nutrient-rich soil patches under ambient light conditions. This inability of *Fagus* is embodied by the significantly reduced root biomass placement under interspecific competition, partially in combination with reduced extracellular enzymatic activities, which resulted in low root N concentrations. Our findings can be generally attributed to the different growth patterns of *Fagus sylvatica* and *Acer pseudoplatanus*, at least at the seedling stage. In studies investigating growth performance, *Acer sp.* had a competitive advantage over *Fagus* with increasing light availability and under non-limiting water supply (e.g., [54,90]).

The observed foraging behavior of *Fagus sylvatica* seedlings under competition seems unfavorable to exploit a specific, nutrient-rich soil patch. However, we speculate that a (potentially resource availability- or kin recognition-induced) feedback mechanism may limit root growth (and related C costs, i.e., respiration, exudation and mycorrhizal symbionts; as evident from our data) into 'pre-occupied' soil areas. Limiting C allocation to specific, 'non-efficient' roots (in regard to resource uptake) may result in an overall greater C use efficiency in *Fagus* seedlings. Diverging C allocation and turnover patterns in specific parts of the root system have been previously shown e.g., in *Pinus sylvestris* [91]. This may foster the ability of highly shade-tolerant *Fagus* seedlings to withstand the limited availability of photosynthetic assimilates—as is prevailing under the light conditions of dense forest understories. Indeed, the importance of C assimilate availability in determining the N uptake capacities of *Acer* and *Fagus* was demonstrated by Simon, Li and Rennenberg [53]—i.e., a reduced light availability severely hampering the ammonium and glutamine uptake of *Acer* but not *Fagus* under interspecific competition. The strongly reduced fine root biomass of *Fagus* might thus be interpreted as a 'self-thinning mechanism', reducing the competitive interactions belowground by curtailing the overlap of root/mycorrhizal zones (and leading to increased root zone 'stratification')—as repeatedly shown for mature *Fagus sylvatica* trees in mixtures (see also theoretical consideration in [38]). While in large parts of Central Europe, the competitive advantage of *Fagus sylvatica* aboveground is clearly related to its ability to tolerate shade in the juvenile state and pre-empt light as mature trees, evidence is thus increasing that *Fagus* 'strength' belowground may not be its competitive effect on neighboring roots per se but its highly plastic and C-efficient root system. To improve our understanding of competitive mechanisms belowground, further studies are needed considering C and nutrient metabolism of roots and their symbionts in relation to whole plant C economics.

Supplementary Materials: The following are available online at http://www.mdpi.com/1999-4907/11/5/528/s1, Table S1: Soil properties in the microcosms and competition chamber. Tables S2–S11: Statistical output of GLMs. Figure S1: Coarse root biomass of *Acer* and *Fagus*. Figure S2: Fine root carbon-to-nitrogen ratio of *Acer* and *Fagus*. Supplementary Material S1: Detailed description of the experimental set-up, irrigation and fertilization, Supplementary Material S2: Methods used to determine soil chemical properties. Supplementary Material S3: Fine root carbon concentration and C:N ratio. Supplementary Material S4: Statistics.

Author Contributions: B.R. and H.S. conceived the experiment. Z.A.L., B.R. and H.S. conducted the experiment and curated and analyzed the data. M.M. calculated the RDPI. Z.A.L. and B.R. wrote the first draft of the paper, Z.A.L., B.R., H.S., D.L.G. and M.M. jointly revised and edited the paper. All authors have read and agreed to the published version of the manuscript.

Funding: Z.L. was partially financially supported by a PhD scholarship (KRG/HCDP program) awarded by the Ministry of Higher Education and Scientific Research, Erbil, Kurdistan Region of Iraq.

Acknowledgments: We thank Melanie Zillinger for her skillful help with the PEA measurements. Judy Simon and three anonymous reviewers are highly acknowledged for critically reviewing earlier versions of the manuscript.

Conflicts of Interest: The authors declare no conflict of interest.

References

1. Coomes, D.A.; Grubb, P.J. Impacts of root competition in forests and woodlands: A theoretical framework and review of experiments. *Ecol. Monogr.* **2000**, *70*, 171–207. [CrossRef]
2. Leuschner, C.; Hertel, D.; Coners, H.; Büttner, V. Root competition between beech and oak: A hypothesis. *Oecologia* **2001**, *126*, 276–284. [CrossRef] [PubMed]
3. Rajaniemi, T.K. Evidence for size asymmetry of belowground competition. *Basic Appl. Ecol.* **2003**, *4*, 239–247. [CrossRef]
4. Weiss, L.; Schalow, L.; Jeltsch, F.; Geissler, K. Experimental evidence for root competition effects on community evenness in one of two phytometer species. *J. Plant Ecol.* **2019**, *12*, 281–291. [CrossRef]
5. Donald, C. The interaction of competition for light and for nutrients. *Aust. J. Agric. Res.* **1958**, *9*, 421–435. [CrossRef]
6. Casper, B.B.; Jackson, R.B. Plant competition underground. *Annu. Rev. Ecol. Syst.* **1997**, *28*, 545–570. [CrossRef]
7. Coomes, D.A.; Allen, R.B. Effects of size, competition and altitude on tree growth. *J. Ecol.* **2007**, *95*, 1084–1097. [CrossRef]
8. De Kroon, H.; Mommer, L.; Nishiwaki, A. Root competition: Towards a mechanistic understanding. In *Root Ecology*; de Kroon, H., Visser, E.J.W., Eds.; Springer: Berlin, Germany, 2003; Volume 168, pp. 215–235.
9. De Kroon, H.; Hendriks, M.; van Ruijven, J.; Ravenek, J.; Padilla, F.M.; Jongejans, E.; Visser, E.J.W.; Mommer, L. Root responses to nutrients and soil biota: Drivers of species coexistence and ecosystem productivity. *J. Ecol.* **2012**, *100*, 6–15. [CrossRef]

10. Trinder, C.J.; Brooker, R.W.; Robinson, D. Plant ecology's guilty little secret: Understanding the dynamics of plant competition. *Funct. Ecol.* **2013**, *27*, 918–929. [CrossRef]
11. Rewald, B.; Razaq, M.; Lixue, Y.; Li, J.; Khan, F.; Jie, Z. Root order-based traits of manchurian walnut & larch and their plasticity under interspecific competition. *Sci. Rep.* **2018**, *8*, 9815.
12. Fowler, N. The role of competition in plant communities in arid and semiarid regions. *Annu. Rev. Ecol. Syst.* **1986**, *17*, 89–110. [CrossRef]
13. Wang, P.; Stieglitz, T.; Zhou, D.W.; Cahill, J.F., Jr. Are competitive effect and response two sides of the same coin, or fundamentally different? *Funct. Ecol.* **2010**, *24*, 196–207. [CrossRef]
14. Chesson, P. Mechanisms of maintenance of species diversity. *Annu. Rev. Ecol. Syst.* **2000**, *31*, 343–366. [CrossRef]
15. Rust, S.; Savill, P.S. The root systems of *Fraxinus excelsior* and *Fagus sylvatica* and their competitive relationships. *Forestry* **2000**, *73*, 499–508. [CrossRef]
16. Büttner, V.; Leuschner, C. Spatial and temporal patterns of fine-root abundance in a mixed oak beech forest. *For. Ecol. Manag.* **1994**, *70*, 11–21. [CrossRef]
17. Rewald, B.; Leuschner, C. Belowground competition in a broad-leaved temperate mixed forest: Pattern analysis and experiments in a four-species stand. *Eur. J. For. Res.* **2009**, *128*, 387–398. [CrossRef]
18. Chave, J.; Coomes, D.; Jansen, S.; Lewis, S.L.; Swenson, N.G.; Zanne, A.E. Towards a worldwide wood economics spectrum. *Ecol. Lett.* **2009**, *12*, 351–366. [CrossRef]
19. Díaz, S.; Kattge, J.; Cornelissen, J.H.C.; Wright, I.J.; Lavorel, S.; Dray, S.; Reu, B.; Kleyer, M.; Wirth, C.; Prentice, I.C. The global spectrum of plant form and function. *Nature* **2016**, *529*, 167–171. [CrossRef]
20. Bardgett, R.D.; Mommer, L.; De Vries, F.T. Going underground: Root traits as drivers of ecosystem processes. *Trends Ecol. Evol.* **2014**, *29*, 692–699. [CrossRef]
21. Reich, P.B. The world-wide 'fast–slow' plant economics spectrum: A traits manifesto. *J. Ecol.* **2014**, *102*, 275–301. [CrossRef]
22. Kunstler, G.; Falster, D.; Coomes, D.A.; Hui, F.; Kooyman, R.M.; Laughlin, D.C.; Poorter, L.; Vanderwel, M.; Vieilledent, G.; Wright, S.J.; et al. Plant functional traits have globally consistent effects on competition. *Nature* **2015**, *529*, 204. [CrossRef] [PubMed]
23. Forey, E.; Langlois, E.; Lapa, G.; Korboulewsky, N.; Robson, T.M.; Aubert, M. Tree species richness induces strong intraspecific variability of beech (*Fagus sylvatica*) leaf traits and alleviates edaphic stress. *Eur. J. For. Res.* **2016**, *135*, 707–717. [CrossRef]
24. Laliberté, E. Below-ground frontiers in trait-based plant ecology. *New Phytol.* **2017**, *213*, 1597–1603. [CrossRef] [PubMed]
25. Kirfel, K.; Heinze, S.; Hertel, D.; Leuschner, C. Effects of bedrock type and soil chemistry on the fine roots of european beech—A study on the belowground plasticity of trees. *For. Ecol. Manag.* **2019**, *444*, 256–268. [CrossRef]
26. Freschet, G.T.; Pagès, L.; Iversen, C.M.; Comas, L.H.; Rewald, B.; Roumet, C.; Klimešová, J.; Zadworny, M.; Poorter, H.; Postma, J.A.; et al. A starting guide to root ecology: Strengthening ecological concepts and standardizing root classification, sampling, processing and trait measurements. *New Phytol.* **2020**. under review.
27. Ostonen, I.; Püttsepp, Ü.; Biel, C.; Alberton, O.; Bakker, M.R.; Lõhmus, K.; Majdi, H.; Metcalfe, D.; Olsthoorn, A.F.M.; Pronk, A.; et al. Specific root length as an indicator of environmental change. *Plant Biosyst.* **2007**, *141*, 426–442. [CrossRef]
28. De la Riva, E.G.; Marañón, T.; Pérez-Ramos, I.M.; Navarro-Fernández, C.M.; Olmo, M.; Villar, R. Root traits across environmental gradients in mediterranean woody communities: Are they aligned along the root economics spectrum? *Plant Soil* **2018**, *424*, 35–48. [CrossRef]
29. Beyer, F.; Hertel, D.; Jung, K.; Fender, A.-C.; Leuschner, C. Competition effects on fine root survival of *Fagus sylvatica* and *Fraxinus excelsior*. *For. Ecol. Manag.* **2013**, *302*, 14–22. [CrossRef]
30. Li, X.; Rennenberg, H.; Simon, J. Competition for nitrogen between *Fagus sylvatica* and *Acer pseudoplatanus* seedlings depends on soil nitrogen availability. *Front. Plant Sci.* **2015**, *6*, 302. [CrossRef]
31. Mayfield, M.M.; Levine, J.M. Opposing effects of competitive exclusion on the phylogenetic structure of communities. *Ecol. Lett.* **2010**, *13*, 1085–1093. [CrossRef]
32. Callaway, R.M.; Pennings, S.C.; Richards, C.L. Phenotypic plasticity and interactions among plants. *Ecology* **2003**, *84*, 1115–1128. [CrossRef]

33. Violle, C.; Enquist, B.J.; McGill, B.J.; Jiang, L.; Albert, C.H.; Hulshof, C.; Jung, V.; Messier, J. The return of the variance: Intraspecific variability in community ecology. *Trends Ecol. Evol.* **2012**, *27*, 244–252. [CrossRef] [PubMed]
34. De Bello, F.; Carmona, C.P.; Mason, N.W.; Sebastià, M.T.; Lepš, J. Which trait dissimilarity for functional diversity: Trait means or trait overlap? *J. Veg. Sci.* **2013**, *24*, 807–819. [CrossRef]
35. Schmid, I.; Kazda, M. Root distribution of Norway spruce in monospecific and mixed stands on different soils. *For. Ecol. Manag.* **2002**, *159*, 37–47. [CrossRef]
36. Weemstra, M.; Mommer, L.; Visser, E.J.W.; van Ruijven, J.; Kuyper, T.W.; Mohren, G.M.J.; Sterck, F.J. Towards a multidimensional root trait framework: A tree root review. *New Phytol.* **2016**, *211*, 1159–1169. [CrossRef]
37. Kong, D.; Wang, J.; Wu, H.; Valverde-Barrantes, O.J.; Wang, R.; Zeng, H.; Kardol, P.; Zhang, H.; Feng, Y. Nonlinearity of root trait relationships and the root economics spectrum. *Nat. Commun.* **2019**, *10*, 2203. [CrossRef]
38. Farrior, C.E. Theory predicts plants grow roots to compete with only their closest neighbours. *Proc. R. Soc. B Biol. Sci.* **2019**, *286*, 20191129. [CrossRef]
39. Faillace, C.A.; Caplan, J.S.; Grabosky, J.C.; Morin, P.J. Beneath it all: Size, not origin, predicts belowground competitive ability in exotic and native shrubs. *J. Torrey Bot. Soc.* **2018**, *145*, 30–40. [CrossRef]
40. Valverde-Barrantes, O.J.; Smemo, K.A.; Feinstein, L.M.; Kershner, M.W.; Blackwood, C.B. Aggregated and complementary: Symmetric proliferation, overyielding, and mass effects explain fine-root biomass in soil patches in a diverse temperate deciduous forest landscape. *New Phytol.* **2015**, *205*, 731–742. [CrossRef]
41. Nicholson, B.A.; Jones, M.D. Early-successional ectomycorrhizal fungi effectively support extracellular enzyme activities and seedling nitrogen accumulation in mature forests. *Mycorrhiza* **2017**, *27*, 247–260. [CrossRef]
42. Paul, C.; Brandl, S.; Friedrich, S.; Falk, W.; Härtl, F.; Knoke, T. Climate change and mixed forests: How do altered survival probabilities impact economically desirable species proportions of Norway spruce and European beech? *Ann. For. Sci.* **2019**, *76*, 14. [CrossRef]
43. Silva Pedro, M.; Rammer, W.; Seidl, R. Tree species diversity mitigates disturbance impacts on the forest carbon cycle. *Oecologia* **2015**, *177*, 619–630. [CrossRef] [PubMed]
44. Petritan, A.M.; Von Lüpke, B.; Petritan, I.C. Effects of shade on growth and mortality of maple (*Acer pseudoplatanus*), ash (*Fraxinus excelsior*) and beech (*Fagus sylvatica*) saplings. *For. Int. J. For. Res.* **2007**, *80*, 397–412. [CrossRef]
45. Tinya, F.; Márialigeti, S.; Bidló, A.; Ódor, P. Environmental drivers of the forest regeneration in temperate mixed forests. *For. Ecol. Manag.* **2019**, *433*, 720–728. [CrossRef]
46. Rewald, B.; Ammer, C.; Hartmann, H.; Malyshev, A.V.; Meier, I.C. Editorial: Woody plants and forest ecosystems in a complex world—Ecological interactions and physiological functioning above and below ground. *Front. Plant Sci.* **2020**. under review. [CrossRef]
47. Ellenberg, H.; Leuschner, C. *Vegetation Mitteleuropas Mit den Alpen*, 6th ed.; Ulmer: Stuttgart, Germany, 2010; p. 1357. (In German)
48. Collet, C.; Fournier, M.; Ningre, F.; Hounzandji, A.P.-I.; Constant, T. Growth and posture control strategies in *Fagus sylvatica* and *Acer pseudoplatanus* saplings in response to canopy disturbance. *Ann. Bot.* **2011**, *107*, 1345–1353. [CrossRef]
49. Brundrett, M.C. Mycorrhizal associations and other means of nutrition of vascular plants: Understanding the global diversity of host plants by resolving conflicting information and developing reliable means of diagnosis. *Plant Soil* **2009**, *320*, 37–77. [CrossRef]
50. Phillips, R.P.; Brzostek, E.; Midgley, M.G. The mycorrhizal-associated nutrient economy: A new framework for predicting carbon-nutrient couplings in temperate forests. *New Phytol.* **2013**, *199*, 41–51. [CrossRef]
51. Lindahl, B.D.; Tunlid, A. Ectomycorrhizal fungi-potential organic matter decomposers, yet not saprotrophs. *New Phytol.* **2015**, *205*, 1443–1447. [CrossRef] [PubMed]
52. Read, D.; Perez-Moreno, J. Mycorrhizas and nutrient cycling in ecosystems—A journey towards relevance? *New Phytol.* **2003**, *157*, 475–492. [CrossRef]
53. Simon, J.; Li, X.; Rennenberg, H. Competition for nitrogen between european beech and sycamore maple shifts in favour of beech with decreasing light availability. *Tree Physiol.* **2014**, *34*, 49–60. [CrossRef] [PubMed]

54. Hommel, R.; Siegwolf, R.; Zavadlav, S.; Arend, M.; Schaub, M.; Galiano, L.; Haeni, M.; Kayler, Z.E.; Gessler, A. Impact of interspecific competition and drought on the allocation of new assimilates in trees. *Plant Biol.* **2016**, *18*, 785–796. [CrossRef] [PubMed]
55. Simon, J.; Dannenmann, M.; Pena, R.; Gessler, A.; Rennenberg, H. Nitrogen nutrition of beech forests in a changing climate: Importance of plant-soil-microbe water, carbon, and nitrogen interactions. *Plant Soil* **2017**, *418*, 89–114. [CrossRef]
56. Lang, C.; Seven, J.; Polle, A. Host preferences and differential contributions of deciduous tree species shape mycorrhizal species richness in a mixed central european forest. *Mycorrhiza* **2011**, *21*, 297–308. [CrossRef]
57. Madsen, P.; Larsen, J.B. Natural regeneration of beech (*Fagus sylvatica* L.) with respect to canopy density, soil moisture and soil carbon content. *For. Ecol. Manag.* **1997**, *97*, 95–105. [CrossRef]
58. Poorter, H.; Bühler, J.; van Dusschoten, D.; Climent, J.; Postma, J.A. Pot size matters: A meta-analysis of the effects of rooting volume on plant growth. *Funct. Plant Biol.* **2012**, *39*, 839–850. [CrossRef]
59. Pritsch, K.; Garbaye, J. Enzyme secretion by ecm fungi and exploitation of mineral nutrients from soil organic matter. *Ann. For. Sci.* **2011**, *68*, 25–32. [CrossRef]
60. Otgonsuren, B.; Rewald, B.; Godbold, D.L.; Göransson, H. Ectomycorrhizal inoculation of *populus nigra* modifies the response of absorptive root respiration and root surface enzyme activity to salinity stress. *Flora Morphol. Distrib. Funct. Ecol. Plants* **2016**, *224*, 123–129. [CrossRef]
61. Wilson, S.D.; Keddy, P.A. Measuring diffuse competition along an environmental gradient—Results from a shoreline plant community. *Am. Nat.* **1986**, *127*, 862–869. [CrossRef]
62. Grace, J.B. On the measurement of plant competition intensity. *Ecology* **1995**, *76*, 305–308. [CrossRef]
63. Valladares, F.; Sanchez-Gomez, D.; Zavala, M.A. Quantitative estimation of phenotypic plasticity: Bridging the gap between the evolutionary concept and its ecological applications. *J. Ecol.* **2006**, *94*, 1103–1116. [CrossRef]
64. R Core Team. *R: A Language and Environment for Statistical Computing*; R Foundation for Statistical Computing: Vienna, Austria, 2019.
65. Ameztegui, A. Plasticity: An R Package to Determine Several Plasticity Indices. 2017. GitHub: GitHub repository. Available online: https://github.com/ameztegui/Plasticity (accessed on 15 April 2019).
66. Moran, M.D. Arguments for rejecting the sequential bonferroni in ecological studies. *Oikos* **2003**, *100*, 403–405. [CrossRef]
67. Kubisch, P.; Hertel, D.; Leuschner, C. Fine root productivity and turnover of ectomycorrhizal and arbuscular mycorrhizal tree species in a temperate broad-leaved mixed forest. *Front. Plant Sci.* **2016**, *7*, 301. [CrossRef] [PubMed]
68. Bolte, A.; Villanueva, I. Interspecific competition impacts on the morphology and distribution of fine roots in european beech (*Fagus sylvatica* L.) and norway spruce (*Picea abies* (L.) karst.). *Eur. J. For. Res.* **2006**, *125*, 15–26. [CrossRef]
69. Zwetsloot, M.J.; Goebel, M.; Paya, A.; Grams, T.E.E.; Bauerle, T.L. Specific spatio-temporal dynamics of absorptive fine roots in response to neighbor species identity in a mixed beech-spruce forest. *Tree Physiol.* **2019**, in press. [CrossRef]
70. Meinen, C.; Hertel, D.; Leuschner, C. Biomass and morphology of fine roots in temperate broad-leaved forests differing in tree species diversity: Is there evidence of below-ground overyielding? *Oecologia* **2009**, *161*, 99–111. [CrossRef]
71. Rewald, B.; Leuschner, C. Does root competition asymmetry increase with water availability? *Plant Ecol. Divers.* **2009**, *2*, 255–264. [CrossRef]
72. Dudley, S.A.; Murphy, G.P.; File, A.L. Kin recognition and competition in plants. *Funct. Ecol.* **2013**, *27*, 898–906. [CrossRef]
73. Hertel, D.; Leuschner, C. The in situ root chamber: A novel tool for the experimental analysis of root competition in forest soils. *Pedobiologia* **2006**, *50*, 217–224. [CrossRef]
74. Fotelli, M.N.; Rudolph, P.; Rennenberg, H.; Geáler, A. Irradiance and temperature affect the competitive interference of blackberry on the physiology of european beech seedlings. *New Phytol.* **2005**, *165*, 453–462. [CrossRef]
75. Simon, J.; Waldhecker, P.; Brüggemann, N.; Rennenberg, H. Competition for nitrogen sources between european beech (*Fagus sylvatica*) and sycamore maple (*Acer pseudoplatanus*) seedlings. *Plant Biol.* **2010**, *12*, 453–458. [CrossRef] [PubMed]

76. Fotelli, M.N.; Rennenberg, H.; Geáler, A. Effects of drought on the competitive interference of an early successional species (*Rubus fruticosus*) on *Fagus sylvatica* L. Seedlings: N-15 uptake and partitioning, responses of amino acids and other n compounds. *Plant Biol.* **2002**, *4*, 311–320. [CrossRef]
77. Rewald, B.; Rechenmacher, A.; Godbold, D.L. It's complicated: Intra-root system variability of respiration and morphological traits in four deciduous tree species. *Plant Physiol.* **2014**, *166*, 736–745. [CrossRef]
78. Meier, I.C.; Angert, A.; Falik, O.; Shelef, O.; Rachmilevitch, S. Increased root oxygen uptake in pea plants responding to non-self neighbors. *Planta* **2013**, *238*, 577–586. [CrossRef] [PubMed]
79. Sandén, H.; Mayer, M.; Stark, S.; Sandén, T.; Nilsson, L.O.; Jepsen, J.U.; Wäli, P.R.; Rewald, B. Moth outbreaks reduce decomposition in subarctic forest soils. *Ecosystems* **2019**, in press.
80. Cheeke, T.E.; Phillips, R.P.; Brzostek, E.R.; Rosling, A.; Bever, J.D.; Fransson, P. Dominant mycorrhizal association of trees alters carbon and nutrient cycling by selecting for microbial groups with distinct enzyme function. *New Phytol.* **2017**, *214*, 432–442. [CrossRef]
81. Guo, Q.; Yan, L.; Korpelainen, H.; Niinemets, Ü.; Li, C. Plant-plant interactions and N fertilization shape soil bacterial and fungal communities. *Soil Biol. Biochem.* **2019**, *128*, 127–138. [CrossRef]
82. Curt, T.; Coll, L.; Prévosto, B.; Balandier, P.; Kunstler, G. Plasticity in growth, biomass allocation and root morphology in beech seedlings as induced by irradiance and herbaceous competition. *Ann. For. Sci.* **2005**, *62*, 51–60. [CrossRef]
83. Bloom, A.J.; Chapin, F.S., III; Mooney, H.A. Resource limitation in plants—An economic analogy. *Annu. Rev. Ecol. Syst.* **1985**, *16*, 363–392. [CrossRef]
84. Bauhus, J.; Messier, C. Soil exploitation strategies of fine roots in different tree species of the southern boreal forest of Eastern Canada. *Can. J. For. Res.* **1999**, *29*, 260–273. [CrossRef]
85. Wang, W.; Wang, Y.; Hoch, G.; Wang, Z.; Gu, J. Linkage of root morphology to anatomy with increasing nitrogen availability in six temperate tree species. *Plant Soil* **2018**, *425*, 189–200. [CrossRef]
86. Lei, P.; Scherer-Lorenzen, M.; Bauhus, J. Belowground facilitation and competition in young tree species mixtures. *For. Ecol. Manag.* **2012**, *265*, 191–200. [CrossRef]
87. McCormack, M.L.; Dickie, I.A.; Eissenstat, D.M.; Fahey, T.J.; Fernandez, C.W.; Guo, D.; Helmisaari, H.S.; Hobbie, E.A.; Iversen, C.M.; Jackson, R.B.; et al. Redefining fine roots improves understanding of belowground contributions to terrestrial biosphere processes. *New Phytol.* **2015**, *207*, 505–518. [CrossRef] [PubMed]
88. Iversen, C.M.; McCormack, M.L.; Powell, A.S.; Blackwood, C.B.; Freschet, G.T.; Kattge, J.; Roumet, C.; Stover, D.B.; Soudzilovskaia, N.A.; Valverde-Barrantes, O.J. A global fine-root ecology database to address below-ground challenges in plant ecology. *New Phytol.* **2017**, *215*, 15–26. [CrossRef]
89. Kattge, J.; DÍAz, S.; Lavorel, S.; Prentice, I.C.; Leadley, P.; BÖNisch, G.; Garnier, E.; Westoby, M.; Reich, P.B.; Wright, I.J.; et al. Try—A global database of plant traits. *Glob. Chang. Biol.* **2011**, *17*, 2905–2935. [CrossRef]
90. Petriţan, A.M.; von Lüpke, B.; Petriţan, I.C. Influence of light availability on growth, leaf morphology and plant architecture of beech (*Fagus sylvatica* L.), maple (*Acer pseudoplatanus* L.) and ash (*Fraxinus excelsior* L.) saplings. *Eur. J. For. Res.* **2009**, *128*, 61–74. [CrossRef]
91. Keel, S.G.; Campbell, C.D.; Högberg, M.N.; Richter, A.; Wild, B.; Zhou, X.; Hurry, V.; Linder, S.; Näsholm, T.; Högberg, P. Allocation of carbon to fine root compounds and their residence times in a boreal forest depend on root size class and season. *New Phytol.* **2012**, *194*, 972–981. [CrossRef]

Article

Mechanical Characteristics of the Fine Roots of Two Broadleaved Tree Species from the Temperate Caspian Hyrcanian Ecoregion

Azade Deljouei [1,2,*], Ehsan Abdi [1], Massimiliano Schwarz [2], Baris Majnounian [1], Hormoz Sohrabi [3] and R. Kasten Dumroese [4]

1 Department of Forestry and Forest Economics, Faculty of Natural Resources, University of Tehran, 31587-77871 Karaj, Iran; abdie@ut.ac.ir (E.A.); bmajnoni@ut.ac.ir (B.M.)

2 School of Agricultural, Forest and Food Sciences HAFL, Bern University of Applied Sciences, 3052 Zollikofen, Switzerland; massimiliano.schwarz@bfh.ch

3 Faculty of Natural Resources, Tarbiat Modares University, 46414-356 Noor, Iran; hsohrabi@modares.ac.ir

4 Rocky Mountain Research Station, Forest Service, U.S. Department of Agriculture, Moscow, ID 83843, USA; kas.dumroese@usda.gov

* Correspondence: a.deljooei@ut.ac.ir

Received: 23 February 2020; Accepted: 17 March 2020; Published: 20 March 2020

Abstract: In view of the important role played by roots against shallow landslides, root tensile force was evaluated for two widespread temperate tree species within the Caspian Hyrcanian Ecoregion, i.e., *Fagus orientalis* L. and *Carpinus betulus* L. Fine roots (0.02 to 7.99 mm) were collected from five trees of each species at three different elevations (400, 950, and 1350 m a.s.l.), across three diameter at breast height (DBH) classes (small = 7.5–32.5 cm, medium = 32.6–57.5 cm, and large =57.6–82.5 cm), and at two slope positions relative to the tree stem (up- and down-slope). In the laboratory, maximum tensile force (N) required to break the root was determined for 2016 roots (56 roots per each of two species x three sites x three DBH classes x two slope positions). ANCOVA was used to test the effects of slope position, DBH, and study site on root tensile force. To obtain the power-law regression coefficients, a nonlinear least square method was used. We found that: 1) root tensile force strongly depends on root size, 2) *F. orientalis* roots are stronger than *C. betulus* ones in the large DBH class, although they are weaker in the medium and small DBH classes, 3) root mechanical resistance is higher upslope than downslope, 4) roots of the trees with larger DBH were the most resistant roots in tension in compare with roots of the medium or small DBH classes, and 5) the root tensile force for both species is notably different from one site to another site. Overall, our findings provide a fundamental contribution to the quantification of the protective effects of forests in the temperate region.

Keywords: bioengineering; *Carpinus betulus*; *Fagus orientalis*; tensile force

1. Introduction

Worldwide, 24 billion tons of fertile soil are estimated to be lost every year due to erosion and mass wasting [1]. Vegetation, especially in the mountain regions, plays a remarkable role in stabilizing slopes and protecting soil from erosion and landslides [2,3]. Particularly for trees, roots are known to reinforce soil through three mechanisms [4–7]: basal root reinforcement, lateral root reinforcement, and increasing the stiffness of the root–soil composite material. In all of these mechanisms, the contribution of roots is defined by their mechanical properties (strength and elasticity) [4,5,8–10] and their density and spatial distribution [4,6,9,11], although species, environment, root diameter, root branching order, age of the trees, root architecture, and forest structure are known to influence root reinforcement variability [12–14].

A review of the literature reveals an overall lack of agreement regarding the size of the roots that must be considered in root mechanical estimation. On one hand, some authors consider only roots with diameters less than 10 mm, because these fine and thin roots act as tensile fibers during slope failures and provide the major contribution to slope stability [15–17]. On the other hand, a few authors indicated that roots of a size up to 20 mm in diameter contribute the most slope stability [18], although, in rare cases, roots with diameter up to 40 mm are said to play an important role in slope stability [19].

In many studies, root reinforcement was estimated for different vegetation species growing in different regions and environments. The mechanical properties of roots have been studied in several works [20,21] and it is known that the mechanical properties of roots change depending on the species and local conditions, such as the amount of nutrients and water content [21–23]. Generally, roots extend close to the soil surface where the soil has the lowest bulk density and water, oxygen, and nutrients are most available. With increasing soil depth, soil bulk density increases and aeration decreases; consequently, root density and size decline [24]. However, a comprehensive and statistically strong analysis of the influence of factors such as region, species, the diameter at breast height (DBH), and slope position of the roots relative to the tree stem on root mechanical characteristics is lacking in the literature, especially in the Hyrcanian Ecoregion. In Iran, this ecoregion is on the UNESCO World Heritage List from July 2019 because of its biological diversity that provides high economic and social value.

The need for more information about root mechanical characteristics is great. For example, Iran ranks within the top 10 countries for high risk of soil erosion and mass soil movement [25]. Annually, the estimated loss of fertile soil in Iran is one to five billion tons per year [26]. From the period 1996 to 2008, Iran recorded 4900 landslides, and even though these landslides were often shallow in nature [25,27], they caused about 30 billion US dollars of damage [25]. Forestry and natural hazards policies need to be based on the scientific background, which can support the decision makers and forest managers. A simple and reliable quantification of the effect of forests on different types of natural hazards is a key step in order to define a comprehensive and feasible risk mitigation strategy. Such issues still remain largely unsolved globally, especially in temperate forests. Vegetation restoration and forest management are important for the mitigation of such phenomena and preventing hazards, mostly through the impact of the reinforcement exerted by root systems [28,29].

Thus, toward providing a fundamental contribution in the quantification afforded by fine roots against a shallow landslide hazard, our study objective was to assess the most important factors that influence tensile force variability within the roots of the most widespread Iranian tree species in the Caspian Hyrcanian Ecoregion. We explored root tensile force in *Carpinus betulus* L. and *Fagus orientalis* L.; these two species were selected because they are the most common species in the Hyrcanian forests and can be found from Europe to the Caucasus and northern Iran [30]. *F. orientalis* is a shade-tolerant species whereas *C. betulus* is semi-shade tolerant [31]. In the Hyrcanian forests of Iran, more than 80 species can be found on typical sites; *F. orientalis* accounts for about 18% of the total forest area, 30% of the standing volume, and 24% of the stem number whereas *C. betulus* contributes 30% of the standing volume and 30% of the total stem number [30].

2. Materials and Methods

2.1. Study Site and Species

Our study site is located in Mazandran Province, northern Iran, within the Hyrcanian Ecoregion along the southern shore of the Caspian Sea. In particular, our sites are in the ~8000 ha Kheyrud Forest (Lat. 36°33′41″ to 36°33′51″ N; Long. 50°33′14″ to 103 50°33′28″ E; WGS84), classified as temperate deciduous forest [31]. Meteorological records from 1961 to 2015 (Nowshahr Meteorological Station: Lat. 36°38′56″ N; Long. 51°29′20″ E; 23 m a.s.l, 7 Km from study area) characterized the mean annual precipitation as 1300 mm, with the heaviest precipitation in fall. October is the wettest month (average 235 mm) and the driest (average 42 mm) is August. The coldest and warmest months are February (7.1 °C) and August (25.1 °C), respectively. The site has a relatively thin soil mantle, and the lithological substrate is mainly calcareous parent material (Jura, Cretaceous) that contains discontinuities and a large number of cracks that can be penetrated by roots [32,33]. The soil in the study site is an Alfisol without any diagnostic horizons [34].

Within the Kheyrud Forest, we collected root samples from three districts (Patom, Namkhane, and Chelir; Figure 1) to compare the effects of different regional environmental factors. These districts, hereafter study sites, range from low elevation (400 m a.s.l.) with the highest mean temperature and lowest annual precipitation (Patom, Lat. 36°36′54″ N; Long. 51°33′49″ E) to mid-elevation (950 m; Namkhane, Lat. 36°33′54″ N; Long. 51°36′09″ E) to the highest elevation (1300 m) having the lowest mean temperature and greatest annual precipitation (Chelir, Lat. 36°32′02″ N; Long. 51°40′24″ E) [35]. These study sites also have different soil properties (Table 1). Soil properties at the three study sites were presented in Table 1 [36]. According to the unified soil classification system [37], the soils on the three study sites were clay with high plasticity (e.g., CH) [36].

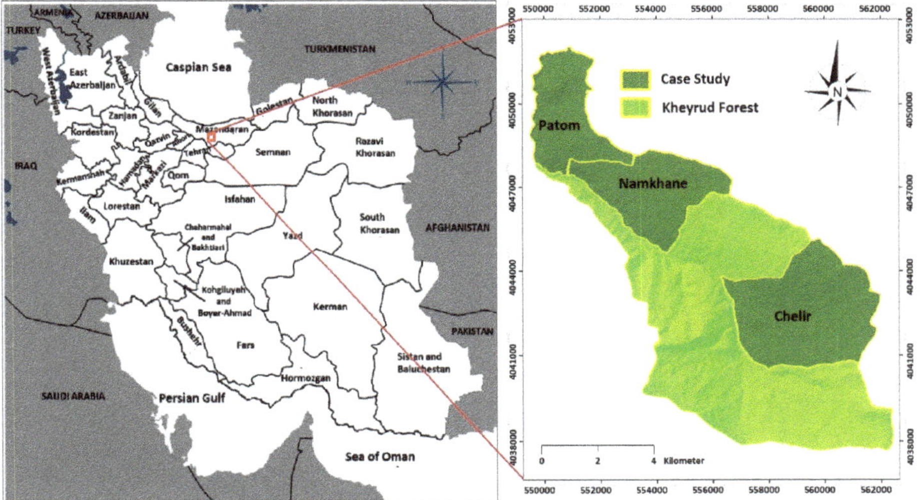

Figure 1. Location of the Chelir, Namkhane, and Patom study sites within the Kheyrud Forest of Mazandran Province in northern Iran. This forest is part of the Caspian Hyrcanian Ecoregion.

2.2. Sampling Design

Three classes of tree diameter at breast height (DBH) (small = 7.5–32.5 cm, medium = 32.6–57.5 cm, and large = 57.6–82.5 cm) were defined to compare the effects of this parameter [39]. On each of the three study sites and for each of three DBH classes, we randomly selected five trees (with at least 6 m distance between trees) for each species (three sites × three DBH classes × five trees × two species = 90 trees total).

Root samples were collected randomly from the soil (0.5 to 1.5 m from the stem) by excavating pits beside the trees at a depth of about 30 cm below the soil surface and from upslope and downslope of the stem [14]. For each species, site, DBH class, and slope position, we collected about 60 root samples (Table 2). At the end of each day of sampling, a 15% alcohol solution was sprayed on roots in order to prevent mould and microbial degradation [17,40], and treated roots were placed into plastic bags and refrigerated (4 °C) until tested; time between sampling and testing in the laboratory was about 48 h.

Table 1. Mean values (± standard error) of soil properties at the three study sites [36].

Soil Properties	Soil Depth (cm)	Study Site		
		Chelir	Namkhane	Patom
Dry density (g cm^{-3})	30	1.13 (± 0.02)	1.09 (± 0.02)	1.16 (± 0.03)
Nitrogen (%)	0–10	22.44 (± 3.02)	24.89 (± 3.02)	21.19 (± 2.25)
	10–20	13.94 (± 1.13)	14.10 (± 0.80)	9.19 (± 0.52)
Phosphorus (ppm)	0–10	20.98 (± 3.01)	25.93 (± 3.13)	23.57 (± 2.11)
	10–20	22.00 (± 2.59)	34.45 (± 1.06)	28.52(± 2.38)
Potassium (ppm)	0–10	1989.2 (± 197.7)	1253.6 (± 131.6)	1068.1 (± 91.6)
	10-20	1228.0 (± 150.5)	1458.3 (± 110.5)	866.7 (± 147.3)
Carbon (%)	0–10	2.84 (± 0.3)	3.12 (± 0.31)	2.50 (± 0.17)
	10–20	1.20 (± 0.10)	1.48 (± 0.13)	0.92 (± 0.06)
Organic matter (%)	0–10	4.90 (± 0.52)	5.38 (± 0.59)	4.31 (± 0.41)
	10–20	2.08 (± 0.19)	2.56 (± 0.25)	1.59 (± 0.14)
pH	0–10	5.54 (± 0.67)	5.25 (± 0.54)	5.42 (± 0.49)
	10–20	5.28 (± 0.57)	5.08 (± 0.36)	5.19 (± 0.40)
EC (ds m^{-1})	0–10	0.45 (± 0.02)	0.45 (± 0.02)	0.38 (± 0.01)
	10–20	0.30 (± 0.01)	0.30 (± 0.01)	0.32 (± 0.01)
Soil liquid limit [1]	30	85.70 (± 6.85)	88.52 (± 7.44)	65 (± 6.21)
Soil plastic limit [1]	30	37.70 (± 3.71)	38.32 (± 4.89)	26.42 (± 3.05)
Soil plasticity index [1]	30	48.00 (± 4.99)	50.20 (± 4.56)	38.58 (± 3.80)
Soil texture [2]		Silt loam	Silt loam	Silt loam
Unified soil classification	-	CH	CH	CH

[1] Atterberg limit [38]. [2] USDA soil classification [34].

Table 2. Total number of root measurements attempted, and in parentheses, the attempts that yielded valid results, for each tree species, diameter at breast height (DBH) class, slope position relative to the main stem, and study site.

Site	DBH Classes [1]	*Carpinus betulus*		*Fagus orientalis*	
		Upslope	Downslope	Upslope	Downslope
Patom	Small	58 (56)	57 (56)	62 (56)	64 (56)
	Medium	59 (56)	59 (56)	61 (56)	62 (56)
	Large	60 (56)	61 (56)	68 (56)	70 (56)
Namkhane	Small	57 (56)	57 (56)	60 (56)	59 (56)
	Medium	58 (56)	60 (56)	61 (56)	62 (56)
	Large	59 (56)	59 (56)	64 (56)	63 (56)
Chelir	Small	58 (56)	59 (56)	58 (56)	57 (56)
	Medium	59 (56)	60 (56)	61 (56)	59 (56)
	Large	62 (56)	61 (56)	63 (56)	62 (56)

[1] DBH classes: small = 7.5–32.5 cm; medium = 32.6–57.5; and large = 57.6–82.5 cm.

Tensile tests were performed using a Universal Testing Machine (SMT-5, SANTOM Co., Tehran, Iran), equipped with 500 kg maximum-capacity load cell (Full Scale, F.S. = accuracy of 0.1% of F.S). Roots with a length of 10 cm and diameters ranging from 0.02 mm to 7.99 mm were clamped into position as vertical as possible within the load cell axis. Roots with diameters more than 8 mm could not be tested due to clumping problems [41]. A strain rate of 10 mm/min [42–44] was applied until breakage occurred; breakage near the middle of the root between the clamps was considered a valid test [45], whereas when breakage occurred proximate to the clamps or breakage was due to slippage or crushing by the clamps, the samples were deemed invalid and discarded (Figure 2). The tensile force (N) was taken as the maximum load at the rupture point. Root diameter was measured at three points near the breaking section [40]. For each site, we obtained 336 valid tests for each species (2016 valid root samples; Table 2).

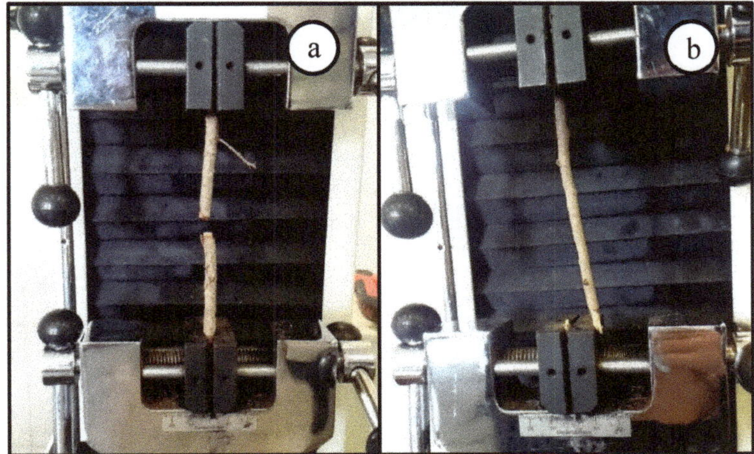

Figure 2. Breaking phase of the root in tensile test. A valid test, where the root broke near the middle (**a**), and an invalid test, where the root broke proximate the clamp (**b**).

2.3. Statistical Analysis

Analysis of covariance (ANCOVA) was used to test the effects of slope position, DBH, and study site on root tensile force. The normality and homogeneity of data were tested before proceeding with the analysis and, as a result, log transformation was necessary to normalize the data. Preliminary use of the ANCOVA revealed that DBH classes and root diameter should be considered as covariate factors, as this model yielded the lowest residuals. Therefore, maximum tensile force was a function of species, slope position, and their interaction. Eta-squared (η^2) was calculated as a measure of the effect of the parameters on the tensile force.

An analysis of residuals was performed to compare the difference between the fitting curves of each dataset as a function of root diameter. The relationships between tensile force–root diameter were interpreted through a regression, which has been proven to be a power-law function [15,40,46]. To obtain the power-law regression coefficients (i.e., F_0 and α), a nonlinear least square method was completed using R software (www.r-project.org, R version 3.3.2, University of Auckland, Auckland, New Zealand). In order to visualize whether or not the differences between datasets are significant, ΔFit and sum of 95% confidence interval (CI) were calculated and compared as shown in Figure 3.

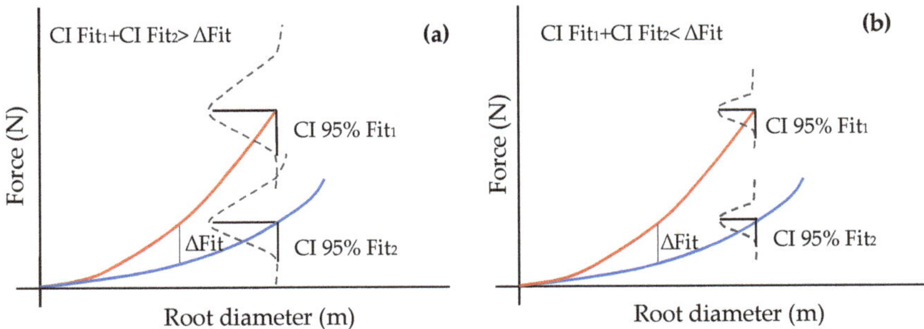

Figure 3. Schematic view of the definition of ΔFit and sum of the confidence interval at 95% (CI-95%), showing an example of overlapping distributions (**a**), and significantly different distributions (**b**).

3. Results

3.1. ANCOVA

A separate covariate analysis for each site revealed that species, slope position, DBH, root diameter, and the interaction of species and slope position were significant ($p < 0.05$) for the tensile force (N) required to break roots (Table 3), except that the interaction was not significant at Namkhane, only at Chelir and Patom. Regardless of site, F values indicate that tensile force was influenced more by root diameter (F=4892) than by species or DBH, and the effect of slope position had the least effect on tensile strength (F=23) (Table 3). The eta-squared (η^2) values indicate that root diameter explains more than 90% of the variance in the studied sites (Table 3).

Table 3. Summary of ANCOVA results for each site, showing the effect of species, slope position, diameter at breast height class (DBH), and the species x slope position interaction on tensile force (N) as a function of root diameter.

Site	Source	df	Sum Square	Mean Square	*F* Value	*P* Value	η^2
Chelir	Species (S)	1	7.63	7.63	388.68	0.000 *	0.049
	Slope position (SP)	1	0.43	0.43	22.15	0.000 *	0.003
	DBH [1]	1	6.13	6.13	312.01	0.000 *	0.039
	Root diameter [1]	1	140.37	140.37	7150.01	0.000 *	0.906
	S × SP	1	0.29	0.29	14.95	0.000 *	0.002
Namkhane	Species (S)	1	2.67	2.67	99.10	0.000 *	0.019
	Slope position (SP)	1	0.40	0.40	14.83	0.000 *	0.003
	DBH [1]	1	0.90	0.90	33.46	0.000 *	0.007
	Root diameter [1]	1	131.95	131.95	4891.88	0.000 *	0.971
	S × SP	1	0.00	0.00	0.02	0.90	0
Patom	Species (S)	1	1.44	1.44	47.86	0.000 *	0.009
	Slope position (SP)	1	0.33	0.33	10.83	0.001 *	0.002
	DBH [1]	1	2.44	2.44	81.41	0.000 *	0.016
	Root diameter [1]	1	146.82	146.82	4891.64	0.000 *	0.970
	S × SP	1	0.26	0.26	8.60	0.003 *	0.002

[1] Data transformed by log 10 to achieve normality. Significant code: '*' <0.05.

3.2. Species

Tensile force values for *F. orientalis* at Chelir, regardless of DBH class, and in the medium DBH class at Patom, were higher than those for *C. betulus* (Figure 4). The slope of regression (F_0) is almost the same for *F. orientalis* and *C. betulus* trees in the smallest DBH class at Namkhane and Patom. In the case of trees with the largest DBH at Namkhane and Patom, as well as those with medium DBH at Namkhane, the tensile force values of *C. betulus* are higher than those of *F. orientalis* (Figure 4). The variability of tensile force within a given species is high because of the wide range of root diameters and different soil properties in these study sites. According to Table 4, significant power-law regressions were observed about the relationship between tensile force and root diameter in all scenarios ($p < 0.05$). For *C. betulus*, F_0 ranges from 13.75 (Chelir site-small DBH) to 76.75 (Namkhane site-medium DBH). The corresponding values for *F. orientalis* were 28.01 (Chelir site -small DBH) and 53.84 (Chelir site-medium DBH) (Table 4). Our results showed that α fluctuates from 1.12 (Namkhane site-large DBH) to 1.41 (Patom site-small DBH) for *C. betulus*. For *F. orientalis*, α variability is higher than that for *C. betulus*, as the highest and lowest α were 0.96 (Patom site-small DBH) and 1.29 (Chelir site-small DBH), respectively (Table 4).

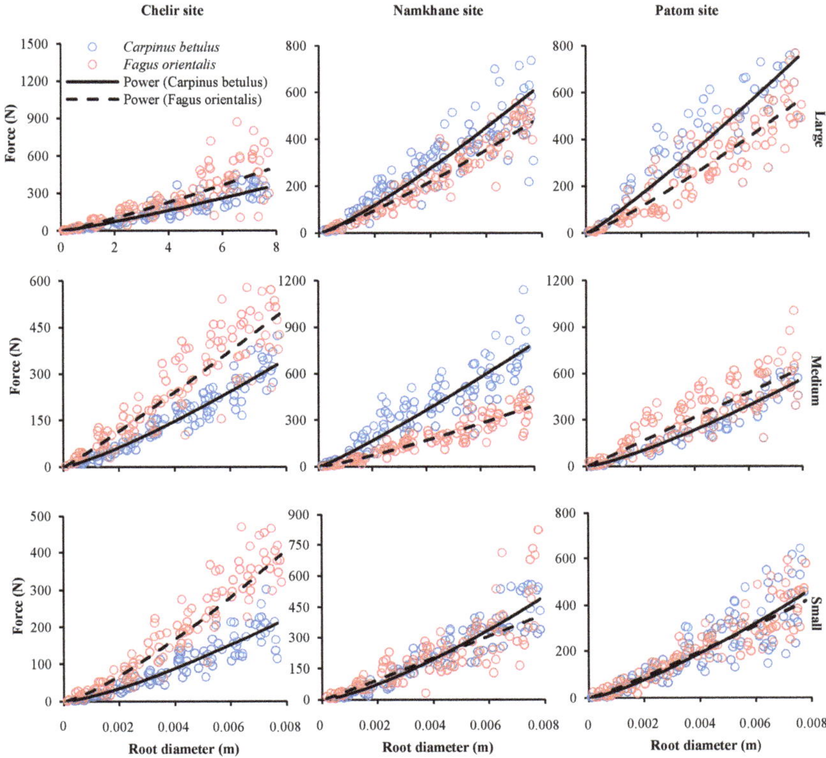

Figure 4. Tensile force at failure point versus root diameter for *Carpinus betulus* (blue circles) and *Fagus orientalis* (red circles) within three classes of diameter at breast height (DBH; small = 7.5–32.5 cm; medium = 32.6–57.5; and large = 57.6–82.5 cm) at three sites (Chelir, Namkhane, Patom). Solid (*C. betulus*) and dashed (*F. orientalis*) lines show power-law regression curves fitted to the data.

Table 4. Coefficients and statistical parameters of the power-law regressions for tensile force–root diameter compared by diameter at breast height (DBH) classes.

Site	Species	DBH [1]	F_0	α	*P* Value	SE
Chelir	*Carpinus betulus*	Large	32.44	1.16	0.000	1.235×10^4
		Medium	26.54	1.24	0.000	1.625×10^4
		Small	13.75	1.34	0.000	2.307×10^4
	Fagus orientalis	Large	44.92	1.17	0.000	7.444×10^4
		Medium	53.84	1.08	0.000	2.224×10^4
		Small	28.01	1.29	0.000	7.143×10^4
Namkhane	*Carpinus betulus*	Large	51.87	1.21	0.000	3.191×10^4
		Medium	76.75	1.12	0.004	5.326×10^4
		Small	28.98	1.37	0.000	1.278×10^5
	Fagus orientalis	Large	42.88	1.18	0.000	5.439×10^4
		Medium	31.51	1.21	0.000	3.827×10^4
		Small	46.42	1.05	0.000	5.166×10^4
Patom	*Carpinus betulus*	Large	75.22	1.33	0.000	2.250×10^4
		Medium	40.21	1.27	0.000	2.190×10^4
		Small	24.25	1.41	0.000	5.236×10^4
	Fagus orientalis	Large	40.66	1.12	0.000	2.528×10^4
		Medium	30.94	1.19	0.000	1.087×10^4
		Small	51.37	0.96	0.000	1.903×10^4

[1] DBH: small = 7.5–32.5 cm; medium = 32.6–57.5; and large = 57.6–82.5 cm.

For tensile force (N), the ANCOVA revealed that at least one species was significantly different on one of the sites (Table 3). With the data for both species combined, we observed that tensile force was not significantly different for trees with large DBH at all three sites and the medium DBH trees at Patom (as evidenced by the overlap of the ΔFit values and sum of the single tail CI-95% values across the range of root diameter classes; Figure 5). Moreover, it was also not significantly different for trees with small DBH at Namkhane and Patom (i.e., ΔFit values are well below the CI-95% values). In contrast, we found a significant difference in tensile force for trees in the small and medium DBH classes at Chelir and in the medium DBH class at Namkhane (i.e., ΔFit values exceed the CI-95% values).

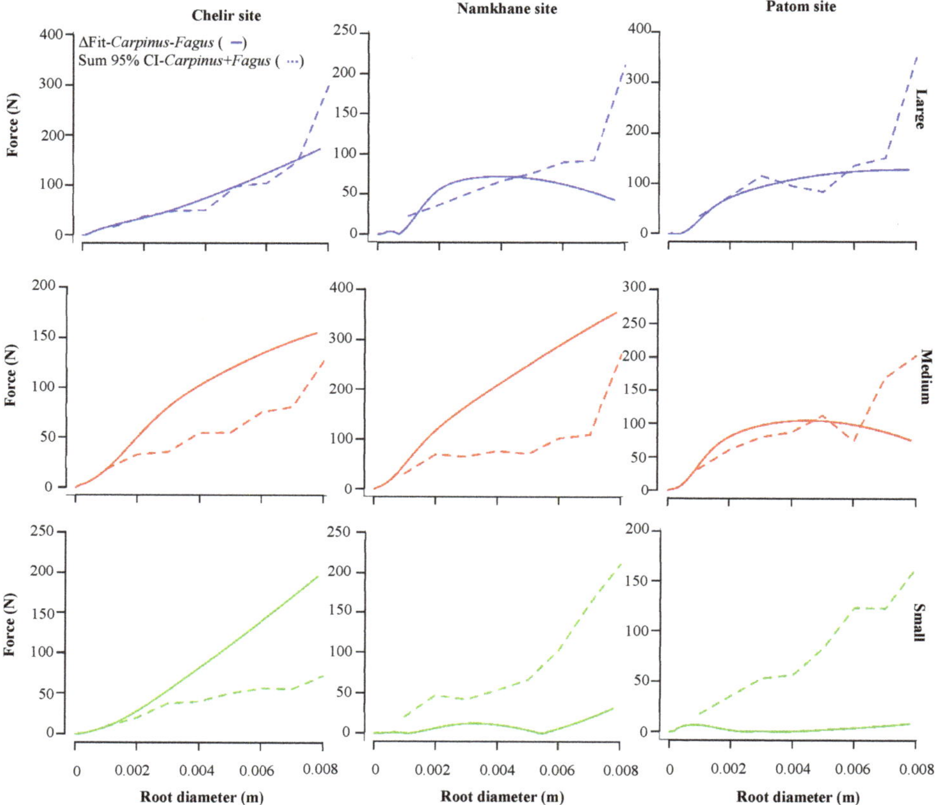

Figure 5. Calculation of difference between the fitting curves of tensile force vs. root diameter for *Carpinus betulus* and *Fagus orientalis* within three classes of diameter at breast height (DBH; small = 7.5–32.5 cm; medium = 32.6–57.5; and large = 57.6–82.5 cm) at three study sites (Chelir, Namkhane, Patom).

3.3. Slope Position

For both species, and across sites and DBH classes, the slope of regression (F_0) for tensile force by increasing root diameter ranged from 19.66 to 47.01 downslope and 26.88 to 54.84 upslope, and α ranged from 1.08 to 1.29 downslope and 1.12 to 1.24 upslope (Table 5). F_0 is slightly higher in the upslope position than in the downslope position at Chelir and Namkhane for *C. betulus* and at Namkhane and Patom for *F. orientalis* (Figure 6 and Table 5). Regression slopes for upslope and downslope positions were similar at Patom for *C. betulus* and for *F. orientalis* at Chelir (Figure 6 and Table 5). Moreover, significant power-law regressions were observed for the relationship between tensile force and root diameter in all scenarios ($p < 0.05$; Table 5). Downslope, α is higher than that for upslope for both species across three study sites, and it ranged from 1.23< α <1.29 and 1.08< α <1.20 for *C. betulus* and *F. orientalis*, respectively (Table 5).

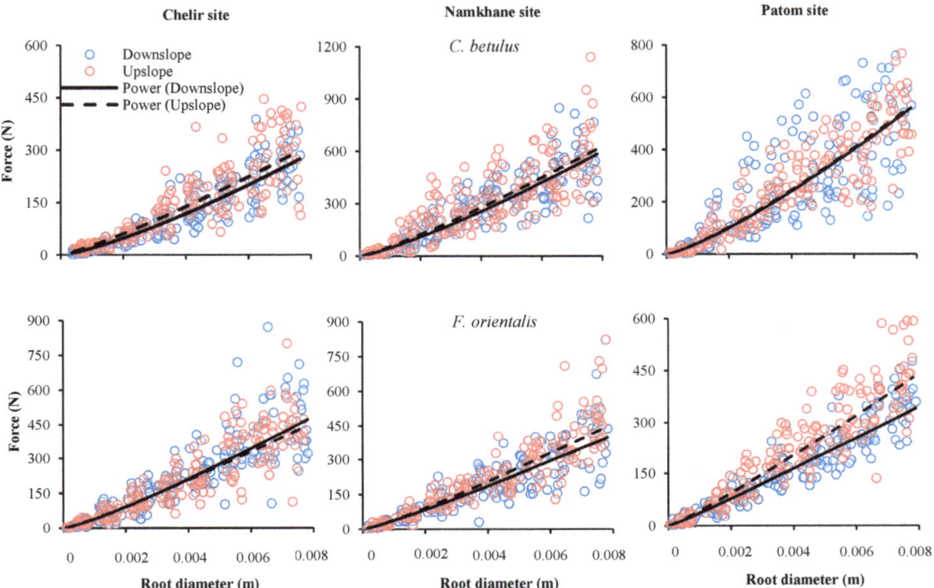

Figure 6. Tensile force at failure point versus root diameter for upslope (red circles) and downslope (blue circles) positions relative to the tree stem at three sites for *Carpinus betulus* and *Fagus orientalis*. Solid (downslope) and dashed (upslope) lines show power-law regression curves fitted to the data.

Table 5. Coefficients and statistical parameters of the power-law regressions for tensile force–root diameter based on a comparison of upslope versus downslope position.

Site	Species	Slope Position	F_0	α	*P* Value	SE
Chelir	*Carpinus betulus*	Down	19.66	1.29	0.000	1.436×10^3
		Up	26.88	1.18	0.000	1.246×10^4
	Fagus orientalis	Down	39.72	1.20	0.000	6.109×10^4
		Up	42.06	1.15	0.000	3.540×10^4
Namkhane	*Carpinus betulus*	Down	47.01	1.23	0.000	3.198×10^4
		Up	54.84	1.17	0.000	4.205×10^4
	Fagus orientalis	Down	38.59	1.13	0.000	5.875×10^4
		Up	41.07	1.16	0.000	4.001×10^4
Patom	*Carpinus betulus*	Down	42.08	1.26	0.000	2.556×10^4
		Up	43.63	1.24	0.000	3.198×10^4
	Fagus orientalis	Down	36.94	1.08	0.000	2.065×10^4
		Up	43.11	1.12	0.000	2.188×10^4

In general, the one-tail confidence intervals (CI) overlap by more than 2.5%, indicating that the differences in the measured maximum tensile forces upslope and downslope were low (Figure 7). For five combinations of species and site (Carpinus-Chelir; Carpinus-Namkhane; Carpinus-Patom; Fagus-Chelir; Fagus-Namkhane), slope position had no significant effect on tensile force by root diameter class (i.e., ΔFit values are well below the CI-95% values) (Figure 7).

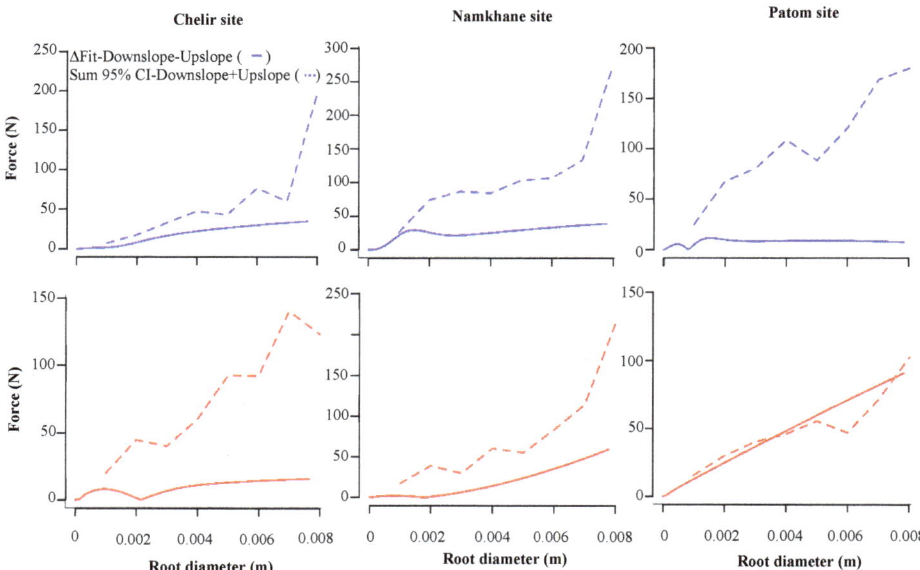

Figure 7. Calculation of difference between the fitting curves of tensile force vs. root diameter for the *Carpinus betulus* (blue line) and *Fagus orientalis* (red line) at three study sites (Chelir, Namkhane, Patom).

3.4. DBH

Significant power-law regressions were found observed for the relationship between tensile force and root diameter for all DBH classess (Table 4). At the three study sites, and for both species, unknown trends were observed for F_0 values; the exception was for small DBH classes of *C. betulus* that had the lowest F_0 values for this species (Table 4). The F_0 of large trees increased slightly at all sites except Namkhane for *C. betulus* and for *F. orientalis* at Chelir (Figure 8). For *C. betulus*, α value within a small DBH class is higher than other DBH classes, whereas no obvious trends were observed for *F. orientalis* (Table 4).

Figure 9 shows the differences between the fitting curves of tensile force vs. root diameter for both species with the three DBH classes on the three study sites. The fitting curves for small and medium, small and large, and medium and large DBH classes were significantly different at all three sites for *C. betulus*, except at Chelir site, where the curves for medium and large were not significantly different (Figure 9). Furthermore, the curves for small and large and medium and large measurements at Namkhane site and small and medium at Patom site overlap for most of the root diameter classes with a probability lower than 2.5% (Figure 9). The fitting curves for small and medium, small and large, and medium and large DBH classes of *F. orientalis* at Chelir and Namkhane sites are not significantly different (Figure 9). Small and medium and small and large measurements at Chelir, and medium and large at Namkhane site, overlap for most of the root diameters, although, in the case of Patom site, ΔFit is higher than sum of single tail CI-95% (Figure 9). Results of the ANCOVA test showed that the effect of DBH classes on tensile force was significant (Table 3). F values of ANCOVA test among DBH classes ranged from 33.46 (Namkhane site) to 312.01 (Chelir site) (Table 3).

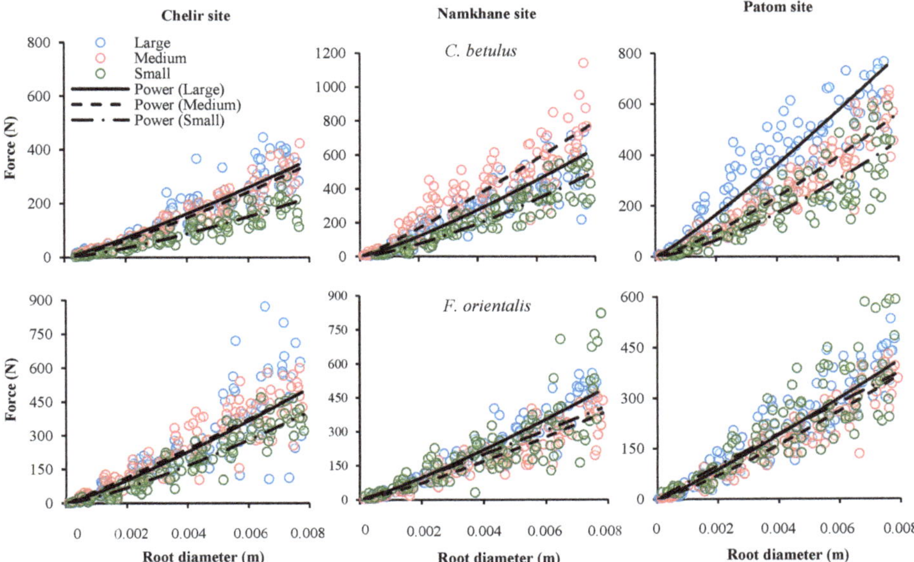

Figure 8. Tensile force at failure point versus root diameter for diameter at breast height (DBH; small = 7.5–32.5 cm; medium = 32.6–57.5; and large = 57.6–82.5 cm) at three study sites (Chelir, Namkhane, Patom) for *Carpinus betulus* and *Fagus orientalis*. Solid (large), dashed (medium), and long dash dot (small) lines show power-law regression curves fitted to the data.

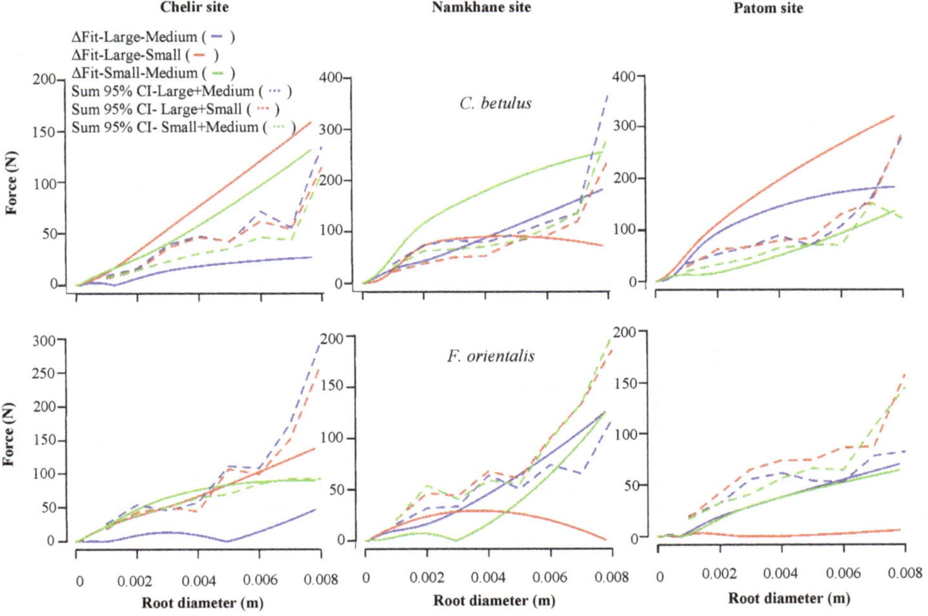

Figure 9. Calculation of difference between the fitting curves of tensile force vs. root diameter for the *Carpinus betulus* and *Fagus orientalis* within three classes of diameter at breast height (DBH; small = 7.5–32.5 cm; medium = 32.6–57.5; and large = 57.6–82.5 cm) at three study sites (Chelir, Namkhane, Patom).

3.5. Study Sites

To evaluate the influence of site on the tensile force–root diameter relationship within each species, data from the different sites were plotted and compared, revealing that tensile force increased slightly at different study sites (Figure 10). Significant power-law regressions were observed about the relationship between tensile force and root diameter across study sites ($p < 0.05$; Table 4). For *C. betulus*, F_0 at Patom site is higher than other sites, while no obvious trends were found for *F. orientalis* (Table 4). For *C. betulus*, the α value observed at the Patom site was greater than that observed on the Chelir and Namkhane sites (Table 4). For *F. orinetalis*, the α value at Namkhane site was greater than that at other sites, except within the small DBH class (Table 4).

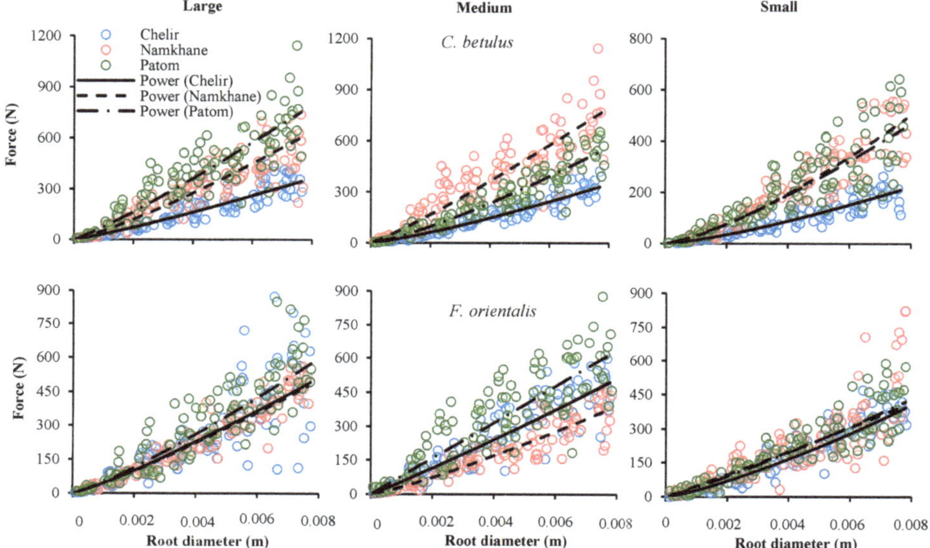

Figure 10. Tensile force at failure point versus root diameter for diameter at breast height (DBH; small = 7.5–32.5 cm; medium = 32.6–57.5; and large = 57.6–82.5 cm) at three study sites (Chelir, Namkhane, Patom). Solid (Chelir), dashed (Namkhane), and long dash dot (Patom) lines show power-law regression curves fitted to the data.

For *C. betulus*, we observed that tensile force was significantly different among Chelir and Namkhane, Chelir and Patom, within all DBH classes (i.e., ΔFit values exceed the CI-95% values), whereas tensile force was not significantly different for large, medium (as evidenced by the overlap of the ΔFit values and sum of the single tail CI-95% values across the range of root diameter classes), and small (i.e., ΔFit values are well below the CI-95% values) trees of Patom and Namkhane (Figure 11). Within the data for *F. orientalis*, tensile force within the medium DBH class at Patom and Namkhane was significant (Figure 11). No significant differences were observed, however, between large and small diameter trees at all sites, medium trees at Chelir and Namkhane, or any trees at Chelir and Patom (Figure 11).

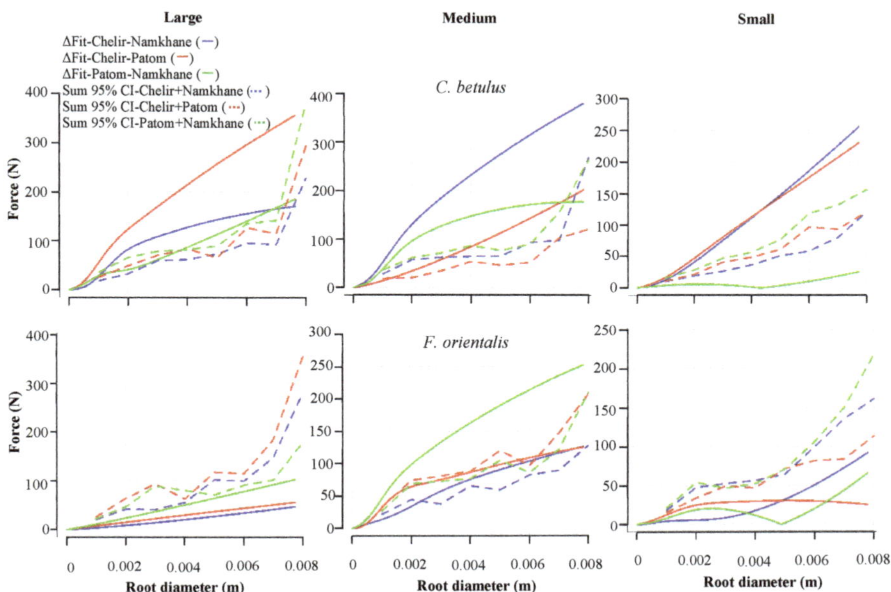

Figure 11. Calculation of difference between the fitting curves of tensile force vs. root diameter of three classes of diameter at breast height (DBH; small = 7.5–32.5 cm; medium = 32.6–57.5; and large = 57.6–82.5 cm) for *Carpinus betulus* and *Fagus orientalis* at three study sites (Chelir, Namkhane, Patom).

4. Discussion

To the best of our knowledge, existing research in the Caspian Hyrcanian Ecoregion on root mechanical characteristics consists of case studies with small datasets focused only on the effects of different species on root mechanical characteristics. The novelty of our research is the broadening of the scope and relevance of root mechanical characteristics. This was achieved through a robust sample of roots (2016 samples) across a variety of DBH classes, elevational gradients, and slope positions of the roots of two common species.

Deciduous broadleaf species characterized by an intensive development of fine roots in the upper soil layer instantly linked to different factors such as climate, age, DBH, and stand composition [47]. Root reinforcement has been noticed as one of the key factors when dealing with slope stability issues and landslides safety, thereby becoming one of the criteria in managing forests against natural hazards [10,11]. Large roots anchor the soil, especially across planes of weakness, and fine roots provide an extensive network that increases soil shear strength [6,48,49]. While coarse roots (>10 mm) have a higher impact on root reinforcement than fine roots [6,7], fine roots are more numerous and occupy a larger area around the tree on slopes than do coarse roots [50], playing a key role in slope stability. Most of the scientists consider that fine and thin roots (<10 mm), which act as tensile fibers during slope failures, provide the major contribution to slope stability [15–17,51]. Large numbers of fine roots could limit the number of cracks occurring on the surface soil, thereby stabilizing the shallow soil more effectively than a small number of coarse roots, which can slip out of the soil upon soil mass sliding [13,52,53].

In older models of root reinforcement, maximum tensile strength of roots was used to describe their mechanical characteristics [54]. Maximum tensile strength is calculated using root diameter and maximum tensile force (the force required to break roots), the later also being calculated in relation to root diameter. Recently, mechanical tensile force has been used independent of tensile strength to describe the mechanical characteristics of roots [7,55]. In our study, our models also focused solely on the tensile force of fine roots to examine their potential contribution to soil reinforcement. We did so to alleviate using the root diameter twice for each observation, which can add uncertainty to the models because of the difficulty involved in accurately determining the very small diameters we studied [40,44]. In this study, roots larger than 8 mm were not tested, therefore, the results of the maximum tensile force will be different if coarse roots are considered in the analysis.

We found a significant relationship between root diameter and tensile force, confirming that root tensile force strongly depends on root size, and similar to the findings of many researchers [15,40,44,46,51,56,57]. This supports the necessity to take the root diameter into account as a covariate for root tensile force analyzing [9,20,40,58]. Moreover, our results show that root tensile force increases with rising root diameter regardless of species, DBH class, or elevation, similar to the results of other studies [6–8,43,57].

In most cases, *F. orientalis*, with higher root resistance than *C. betulus*, may be preferred in soil bioengineering systems. Perhaps this type of effect is related to species-specific tree longevity. Alidadi [59] showed that *F. orientalis* had a significantly higher longevity than *C. betulus*, and Abdi [60] observed that *F. orientalis* roots were stronger than those of *C. betulus* of unspecified DBH in the Kheyrud forest. Our results, which included discrete DBH classes, were similar. Moreover, our comparisons of tensile force vs. root diameter for these two species are similar to those observed in other studies (*F. sylvatica* [22,40,61]; *C. betulus* [40]).

We did not, however, note any differences in root-system mechanical resistance on the basis of slope position. This contrasts with the results of Stokes [62], where resistance was greater in the upslope position compared to the downslope position but concurs with Khuder et al. [63] and Genet et al. [64], who observed that no differences occur in the mechanical properties of the roots in different positions around the root system. Commonly, it is believed that no general rule exists to explain the differences in root growth and mechanical properties on the slope, similar to our results. Although earlier studies examined a variety of parameters on root tensile force (e.g., slope position, species, soil types) [9,10,42,51,53,56,58,62]; to the best of our knowledge, we are the first to attempt to examine the interactions of these parameters. Our finding of an interaction between species and slope positions on two of three sites (i.e., Chelir and Patom) suggests that the effective parameters on root tensile force are various and complex.

In general, we found that the roots of the trees with larger DBH were more resistant in tension compared with the roots of the medium or small DBH classes, which concurs with [11]. This may be explained by the report of [65], who showed that root resistance was lower in the early growth stage and increased in older plants.

For the same root diameter, the force needed to break a root was found to decrease with increasing elevation above sea level [64]. Whenever the tensile strength of the roots decreases, the root system may adapt based on the situation, and therefore tree anchorage is not comprised [64]. Adaption of the root system architecture to external stimuli in response to mechanical force was recently shown as the initiation of new coarse roots [66]. This potential of adaptation based on the environmental conditions is termed "morphoplasticity" [67] or "phenotypic plasticity" and results in asymmetric root distribution [66–70]. Our results showed that for *C. betulus* and *F. orientalis* the root tensile force is notably different from one site to another site, although further assessment on root system architecture is necessary to determine whether any decrease in root resistance associated with increasing elevation was compensated by morphological adaptation. These findings are similar to those of other studies performed on a single species [64] or in a small area [71]. These authors attributed site differences to elevation and the position of the sampling on hillslope. Genet et al. [65], found a significant difference for the same species sampled at different locations, but the elevation between their locations was much greater than those in our study. However, in contrast with the results of Genet et al. [65], the maximum tensile force of roots of *Castanea sativa* Mill. did not differ between the two sites under consideration [7]. The mechanical properties of roots between sites with different soil water content regimes differed significantly [23]. Vergani et al. [40] clarified that the comparisons of sites' relationships for different species remarkably are a consequence of varying conditions of the growing sites and environment. It is discussed that alterations in root mechanical resistance with increasing elevation were due to modification in soil's chemical and physical properties [64]. On our sites, however, the physical characteristics are almost the same in different sites, although chemical properties showed differences in nitrogen, phosphorus, potassium, carbon, nitrogen, and organic matter (Table 1), making it difficult to reach any conclusion. Different levels of soil compaction did not significantly affect the root resistance in the woody plants *Acacia senegal* L. and *Prosopis juliflora* DC. [72]. Goodman and Ennos [73] demonstrated, however, that the roots of two annual plants, *Zea mays* L. and *Helianthus annuus* L., either became stiffer or were unchanged, respectively, when growing in soil with low bulk density Moreover, Jourgholami et al. [74] investigated the ratio of lateral to main root length, which is significantly reduced in high intensity compaction in comparison with control treatment, although the ratio of lateral to main root biomass was not significant among different soil compaction treatments. Accordingly, to prove if soil physical characteristics affect root resistance and mechanical properties, further studies are required.

The increased tensile force with increasing root diameter, slope position, tree age and longevity, and elevation has been attributed to a greater cellulose content as a function of root size, environmental stimuli, or genetics [22,41,42,65,71,75,76]. Consequently, analyzing the cellulose content of roots for different size, DBH, slope position, and age classes would be a valuable addition to the literature.

5. Conclusions

This study provides a comprehensive analysis of root tensile force variation of two common species, *Carpinus betulus* and *Fagus orientalis*, across differing DBH classifications, slope positions, and study site elevations within a temperate deciduous forest. Our results identified that the main factors affecting variability in fine root tensile force are the tree species and the DBH of the sampled tree. Study site location had an effect on root tensile force only for *C. betulus*. No differences in tensile force were found for roots growing upslope or downslope of the tree stem. This information is useful for scientists and forest land managers in order to evaluate the variability in root reinforcement due to different factors, and to account for this uncertainty when evaluating the effectiveness of slope stabilization using biological engineering measures. The selection of a species in a region can cause changes in soil reinforcement and slope stability. Understanding the relationship between tree species and soil bioengineering and the impact of different DBH classes on these processes can be useful for forest management and the selection of appropriate species for reforestation projects. In the Caspian Hyrcanian Ecoregion, *F. orientalis*, with a higher fine root resistance than *C. betulus*, may be preferred in soil bioengineering systems.

Author Contributions: Conceptualization, A.D., E.A. and M.S.; methodology, A.D. and M.S.; software, A.D., M.S. and H.S.; formal analysis, A.D.; data curation, A.D., H.S. and M.S.; writing—original draft preparation, A.D., E.A., M.S. and R.K.D.; writing—review and editing, R.K.D.; supervision, E.A. and B.M.; project administration, E.A. and B.M.; funding acquisition, E.A. and R.K.D. All authors have read and agreed to the published version of the manuscript.

Funding: This research was funded by Iran National Science Foundation (INSF), grant number 93022486.

Acknowledgments: The research was funded by the University of Tehran. The authors express their sincere appreciation to Seyed Mohammad Moein Sadeghi, Rahmatollah Ghomi, Habib Zalnezhad, Ghodrat Daneshvar, Nasrollah Ebrahimpour, and Gholi Kalantaj who assited with field work. The authors also thank Seyed Mohammad Moein Sadeghi, Soghra Keybondori, Fardin Moradi, Fardin Moradzadeh, and Abasalt Tarverdi for laboratory support. Additional support was provided by the USDA Forest Service, National Center for Reforestation, Nurseries, and Genetic Resources and the Rocky Mountain Research Station. The views expressed are strictly those of the authors and do not necessarily represent the positions or policy of their respective institutions.

Conflicts of Interest: The authors declare no conflict of interest.

References

1. Mickovski, S.B.; Gonzalez-Ollauri, A.; Tardio, G. Novel approaches to quantification of the vegetation effects on soil strength. In Proceedings of the 19th International Conference on Soil Mechanics and Geotechnical Engineering, Seoul, Korea, 17–22 September 2017.

2. Schwarz, M.; Phillips, C.; Marden, M.; McIvor, I.R.; Douglas, G.B.; Watson, A. Modelling of root reinforcement and erosion control by Veronese poplar on pastoral hill country in New Zealand. *N. Z. J. For. Sci.* **2016**, *46*, 4. [CrossRef]
3. Sidle, R.C.; Bogaard, T.A. Dynamic earth system and ecological controls of rainfall-initiated landslides. *Earth-Sci. Rev.* **2016**, *159*, 275–291. [CrossRef]
4. Stokes, A.; Atger, C.; Bengough, A.G.; Fourcaud, T.; Sidle, R.C. Desirable plant root traits for protecting natural and engineered slopes against landslides. *Plant Soil* **2009**, *324*, 1–30. [CrossRef]
5. Schwarz, M.; Rist, A.; Cohen, D.; Giadrossich, F.; Egorov, P.; Buttner, D.; Stolz, M.; Thormann, J.J. Root reinforcement of soils under compression. *J. Geophys. Res. Earth Surf.* **2015**, *120*, 2103–2120. [CrossRef]
6. Vergani, C.; Giadrossich, F.; Buckley, P.; Conedera, M.; Pividori, M.; Salbitano, F.; Rauch, H.S.; Lovreglio, R.; Schwarz, M. Root reinforcement dynamics of European coppice woodlands and their effect on shallow landslides: A review. *Earth-Sci. Rev.* **2017**, *167*, 88–102. [CrossRef]
7. Dazio, E.; Conedera, M.; Schwarz, M. Impact of different chestnut coppice managements on root reinforcement and shallow landslide susceptibility. *For. Ecol. Manag.* **2018**, *417*, 63–76. [CrossRef]
8. Schwarz, M.; Giadrossich, F.; Cohen, D. Modeling root reinforcement using a root failure weibull survival function. *Hydrol. Earth Syst. Sci.* **2013**, *17*, 4367–4377. [CrossRef]
9. Moresi, F.V.; Maesano, M.; Matteucci, G.; Romagnoli, M.; Sidle, R.C.; Scarascia Mugnozza, G. Root biomechanical traits in a montane Mediterranean forest watershed: Variations with species diversity and soil depth. *Forests* **2019**, *10*, 341. [CrossRef]
10. Tsige, D.; Senadheera, S.; Talema, A. Stability analysis of plant-root-reinforced shallow slopes along mountainous road corridors based on numerical modeling. *Geosciences* **2020**, *10*, 19. [CrossRef]
11. Cohen, D.; Schwarz, M. Tree-root control of shallow landslides. *Earth Surf. Dyn.* **2017**, *5*, 451–477. [CrossRef]
12. Roering, J.; Schmidt, K.M.; Stock, J.D.; Dietrich, W.E.; Montgomery, D.R. Shallow landsliding, root reinforcement, and the spatial distribution of tress in the Oregon Coast range. *Can. Geotech. J.* **2003**, *40*, 237–253. [CrossRef]
13. Loades, K.; Bengough, A.; Bransby, M.; Hallett, P. Planting density influence on fibrous root reinforcement of soils. *Ecol. Eng.* **2010**, *36*, 276–284. [CrossRef]
14. Mao, Z.; Wang, Y.; McCormack, M.L.; Rowe, N.; Deng, X.; Yang, X.; Xia, S.; Nespoulous, J.; Sidle, R.C.; Guo, D.; et al. Mechanical traits of fine roots as a function of topology and anatomy. *Ann. Bot* **2018**, *122*, 1103–1116. [CrossRef]
15. Genet, M.; Stokes, A.; Fourcaud, T.; Norris, J.E. The influence of plant diversity on slope stability in a moist evergreen deciduous forest. *Ecol. Eng.* **2010**, *36*, 265–275. [CrossRef]
16. Abdi, E.; Majnounian, B.; Genet, M.; Rahimi, H. Quantifying the effects of root reinforcement of Persian ironwood (*Parrotia persica*) on slope stability; a case study: Hillslope of Hyrcanian forests, Northern Iran. *Ecol. Eng.* **2010**, *36*, 1409–1416. [CrossRef]
17. Bassanelli, C.; Bischetti, G.B.; Chiaradia, E.A.; Rossi, L.; Vergani, C. The contribution of chestnut coppice forests on slope stability in abandoned territory: A case study. *JAE* **2013**, *44*, 68–73. [CrossRef]
18. Styczen, M.E.; Morgan, R.P.C. *Engineering Properties of Vegetation*; Taylor & Francis: London, UK, 1995.
19. O'Loughlin, C.L.; Watson, A. Root-wood strength deterioration in radiata pine after clearfelling. *N. Z. J. For. Sci.* **1979**, *9*, 284–293.
20. Hales, T.; Cole-Hawthorne, C.; Lovell, L.; Evans, S.L. Assessing the accuracy of simple field based root strength measurements. *Plant Soil* **2013**, *372*, 553–565. [CrossRef]
21. Zhang, C.B.; Chen, L.H.; Jiang, J. Why fine tree roots are stronger than thicker roots: The role of cellulose and lignin in relation to slope stability. *Geomorphology* **2014**, *206*, 196–202. [CrossRef]
22. Genet, M.; Stokes, A.; Salin, F.; Mickovski, S.B.; Fourcaud, T.; Dumail, J.F.; Van Beek, R. The influence of cellulose content on tensile strength in tree roots. *Plant Soil* **2005**, *278*, 1–9. [CrossRef]
23. Hales, T.C.; Miniat, C.F. Soil moisture causes dynamic adjustments to root reinforcement that reduce slope stability. *Earth Surf. Process. Landforms.* **2017**, *42*, 803–817. [CrossRef]
24. Dobson, M. Tree root system. *Arboric* **1995**, *130*, 1–6.
25. Yarahmadi, J.; Rostaei, S.H.; Sharifikia, M.; Rostaei, M. Identifying and monitoring slope instability by using Differential SAR Interferometry, case study: Garmichay watershed, Miyane. *JQGR* **2015**, *3*, 44–59.
26. Khajavi, E.; Arabkhedri, M.; Mahdian, M.H.; Shadfar, S. Investigation of Water Erosion and Soil Loss Values with using the Measured Data from Cs-137 Method and Experimental Plots in Iran. *JWMR* **2015**, *6*, 137–151.

27. Karam, A. Quantitive Modelling and Landslide Risk Zonation in Faulted Zageros, Case Study: Sarkhon Basin in Char-Mahal-e-Bakhteyari. Ph.D. Thesis, Tarbiat Modares University, Noor, Iran, 2001.
28. Sakals, M.E.; Sidle, R.C. A spatial and temporal model of root cohesion in forest soils. *Can. J. For. Res.* **2004**, *34*, 950–958. [CrossRef]
29. Burylo, M.; Hudek, C.; Rey, F. Soil reinforcement by the roots of six dominant species on eroded mountainous marly slopes (Southern Alps, France). *Catena* **2011**, *84*, 70–78. [CrossRef]
30. Sagheb-Talebi, K.; Sajedi, T.; Pourhashemi, M. *Forests of Iran: A Treasure from the Past, a Hope for the Future*; Springer: Berlin/Heidelberg, Germany, 2014.
31. Marvie-Mohadjer, M.R. *Silviculture*, 5th ed.; University of Tehran Press: Tehran, Iran, 2019; p. 418.
32. Deljouei, A.; Abdi, E.; Marcantonio, M.; Majnounian, B.; Amici, V.; Sohrabi, H. The impact of forest roads on understory plant diversity in temperate hornbeam-beech forests of Northern Iran. *Environ. Monit. Assess.* **2017**, *189*, 392. [CrossRef]
33. Deljouei, A.; Sadeghi, S.M.M.; Abdi, E.; Bernhardt-Römermann, M.; Pascoe, E.L.; Marcantonio, M. The impact of road disturbance on vegetation and soil properties in a beech stand, Hyrcanian forest. *Eur. J. For. Res.* **2018**, *137*, 759–770. [CrossRef]
34. Soil Survey Staff. *Soil Characterization and Profile Description Data*; Soil Survey Laboratory, Natural Resources Conservation Service, USDA: Lincoln, NE, USA, 1995.
35. Azaryan, M.; Marvie-Mohadjer, M.R.; Etemaad, V.; Shirvany, A.; Sadeghi, S.M.M. Morphological characteristics of old trees in Hyrcanian forest (Case study: Pattom and Namkhaneh districts, Kheyrud). *JFWP* **2015**, *68*, 47–59.
36. Deljouei, A. Spatial Dynamics of Soil Reinforcement Due to Presence of Roots in Hyrcanian Forest (Case Study: Kheyroud Forest). Ph.D. Thesis, University of Tehran, Karaj, Iran, 2019.
37. ASTM. ASTM D4318-00. In *Standard Test Methods for Liquid Limit, Plastic Limit, and Plasticity Index of Soils, Annual Book of ASTM Standards*; American Society for Testing and Materials: West Conshohocken, PA, USA, 2003; Volume 04.08, pp. 582–595.
38. Anonymous. Technical Memo No. 3-357. In *The Unified Soil Classification System*; United States Army Corps of Engineers: Vicksburg, MS, USA, 1957.
39. Sagheb-Talebi, K.; Eslami, A. Nature-based silviculture—How can we achieve the equilibrium state in uneven-aged oriental beech stands? In Proceedings of the 8th International Symposium, Hokkaido, Japan, 8–13 September 2008.
40. Vergani, C.; Chiaradia, E.; Bischetti, G. Variability in the tensile resistance of roots in alpine forest tree species. *Ecol. Eng.* **2012**, *46*, 43–56. [CrossRef]
41. De Baets, S.; Torri, D.; Poesen, J.; Meersmans, J. Modelling increased soil cohesion due to roots with EUROSEM. *Earth Surf. Process. Landforms.* **2008**, *33*, 1948–1963. [CrossRef]
42. Chiaradia, E.A.; Vergani, C.; Bischetti, G.B. Evaluation of the effects of three European forest types on slope stability by field and probabilistic analyses and their implications for forest management. *For. Ecol. Manag.* **2016**, *370*, 114–129. [CrossRef]
43. Vergani, C.; Schwarz, M.; Soldati, M.; Corda, A.; Giadrossich, F.; Chiaradia, E.A.; Morando, P.; Bassanelli, C. Root reinforcement dynamics in subalpine spruce forests following timber harvest: A case study in Canton Schwyz, Switzerland. *Catena* **2016**, *143*, 257–288. [CrossRef]
44. Gilardelli, F.; Vergani, C.; Gentili, R.; Bonis, A.; Pierre, C.; Sandra, C.; Chiaradia, E.A. Root characteristics of herbaceous species for topsoil stabilization restoration in projects. *Land Degrad. Dev.* **2017**, *28*, 2074–2085.
45. Abdi, E. Effect of oriental beech root reinforcement on slope stability (Hyrcanian Forest, Iran). *J. For. Sci.* **2014**, *60*, 166–173. [CrossRef]
46. Nilaweera, N.; Nutalaya, P. Role of tree roots in slope stabilisation. *Bull. Eng. Geol. Environ.* **1999**, *57*, 337–342. [CrossRef]
47. Yuan, Z.Y.; Chen, H.Y.H. Fine root biomass, production, turnover rates, and nutrient contents in boreal forest ecosystems in relation to species, climate, fertility, and stand age: Literature review and meta-analyses. *Crit. Rev. Plant Sci.* **2010**, *29*, 204–221. [CrossRef]
48. Sidle, R.C.; Ochiai, H. *Landslides: Processes, Prediction and Land Use*; American Geographical Union: Washington, DC, USA, 2006.
49. Forbes, K.; Broadhead, J. Forest and landslides. In *The Role of Trees and Forests in the Prevention of Landslides and Rehabilitation of Landslide-Affected Areas in Asia*; RAP Publication: Bangkok, Thailand, 2013.

50. Schwarz, M.; Cohen, D.; Or, D. Root-soil mechanical interactions during pullout and failure of root bundles. *J. Geophys. Res.* **2010**, *115*, F4. [CrossRef]
51. Abdi, E. Root tensile force and resistance of several tree and shrub species of Hyrcanian forest, Iran. *Croat. J. For. Eng.* **2018**, *39*, 255–270.
52. Comino, E.; Marengo, P.; Rolli, V. Root reinforcement effect of different grass species: A comparison between experimental and models results. *Soil Tillage Res.* **2010**, *110*, 60–68. [CrossRef]
53. Lee, J.T.; Chu, M.Y.; Lin, Y.S.; Kung, K.N.; Lin, W.C.; Lee, M.J. Root traits and biomechanical properties of three tropical pioneer tree species for forest restoration in landslide areas. *Forests* **2020**, *11*, 179. [CrossRef]
54. Wu, T.H. No 5, Dpt. of Civil Engineering. In *Investigation on landslides on Prince of Wales Island, Alaska Geotech Rpt*; Ohio State University: Columbus, OH, USA, 1976.
55. Schwarz, M.; Cohen, D.; Or, D. Spatial characterization of root reinforcement at stand scale: Theory and case study. *Geomorphology* **2012**, *171*, 190–200. [CrossRef]
56. Abdi, E.; Deljouei, A. Seasonal and spatial variability of root reinforcement in three pioneer species of the Hyrcanian forest. *Austrian J. For. Sci.* **2019**, *136*, 175–198.
57. Boldrin, D.; Leung, A.K.; Bengough, A.G. Root biomechanical properties during establishment of woody perennials. *Ecol. Eng.* **2017**, *109*, 196–206. [CrossRef]
58. Abdi, E.; Saleh, H.R.; Majnounian, B.; Deljouei, A. Soil fixation and erosion control by *Haloxylon persicum* roots in arid lands, Iran. *J. Arid. Land.* **2019**, *11*, 86–96. [CrossRef]
59. Alidadi, F. Decay Dynamic of Beech and Hornbeam Dead Trees in Mixed Beech (*Fagus orientalis* Lipsky) Stands. Master's Thesis, University of Tehran, Karaj, Iran, 2014.
60. Abdi, E. An Investigation of the Effect of Tree Roots in Slope Stability in Order to Use in Practical Forest Road Construction and Bioengineering. Ph.D. Thesis, University of Tehran, Karaj, Iran, 2009.
61. Bischetti, G.B.; Chiaradia, E.A.; Epis, T.; Morlotti, E. Root cohesion of forest species in the Italian Alps. *Plant Soil* **2009**, *324*, 71–89. [CrossRef]
62. Stokes, A. Biomechanics of tree root anchorage. In *Plant Roots: The Hidden Half*, 3rd ed.; Waisel, Y., Eshel, A., Kafkafi, U., Eds.; CRC Press: Boca Raton, FL, USA, 2002; pp. 175–186.
63. Khuder, H.; Danjon, F.; Stokes, A.; Fourcaud, T. Growth response and root architecture of black locust seedlings growing on slopes and subjected to mechanical perturbation. In Proceedings of the 5th plant biomechanics conference, Stockholm, Sweden, 28 August–1 September 2006.
64. Genet, M.; Li, M.; Luo, T.; Fourcaud, T.; Clément-Vidal, A.; Stokes, A. Linking carbon supply to root cell-wall chemistry and mechanics at high altitudes in *Abies Georgei*. *Ann. Bot.* **2011**, *107*, 311–320. [CrossRef]
65. Genet, M.; Stokes, A.; Fourcaud, T.; Cai, X.; Lu, Y. Soil fixation by tree roots: Changes in root reinforcement parameters with age in *Cryptomeria Japonica* D. Don. In *Plantations. Disaster Mitigation of Debris Flows, Slope Failures and Landslides*; Universal Academy Press, Inc.: Tokyo, Japan, 2006; pp. 535–542.
66. Montagnoli, A.; Terzaghi, M.; Chiatante, D.; Scippa, G.S.; Lasserre, B.; Dumroese, R.K. Ongoing modifications to root system architecture of *Pinus ponderosa* growing on a sloped site revealed by tree-ring analysis. *Dendrochronologia* **2019**, *58*, 125650. [CrossRef]
67. Marler, T.E.; Discekici, H.M. Root development of 'Red Lady' papaya plants grown on a hillside. *Plant Soil* **1997**, *195*, 37–42. [CrossRef]
68. Di Iorio, A.; Lasserre, B.; Scippa, G.S.; Chiatante, D. Root system architecture of *Quercus pubescens* trees growing on different sloping conditions. *Ann. Bot.* **2005**, *95*, 351–361. [CrossRef]
69. Ganatsas, P.; Spanos, I. Root system asymmetry of Mediterranean pines. *Plant Soil* **2005**, *278*, 75–83. [CrossRef]
70. Dumroese, R.K.; Terzaghi, M.; Chiatante, D.; Scippa, G.S.; Lasserre, B.; Montagnoli, A. Functional traits of *Pinus ponderosa* coarse-roots in response to slope conditions. *Front. Plant Sci.* **2019**, *10*, 947. [CrossRef]
71. Hales, T.C.; Ford, C.R.; Hwang, T.; Vose, J.M.; Band, L.E. Topographic and ecologic controls on root reinforcement. *J. Geophys. Res. Earth Surf.* **2009**, *114*, F03013. [CrossRef]
72. Ba, M. Etude des Proprietes Biomecaniques et de La Capacite de Vie Symbiotique de Racines D'arbres D'acacia Senegal Willd et de *Prosopis juliflora* DC. Ph.D. Thesis, University Bordeaux, Bordeaux, France, 2008.
73. Goodman, A.M.; Ennos, A.R. The effects of soil bulk density on the morphology and anchor age mechanics of the root systems of sunflower and maize. *Ann. Bot.* **1999**, *83*, 293–302. [CrossRef]
74. Jourgholami, M.; Khoramizadeh, A.; Zenner, E.K. Effects of soil compaction on seedling morphology, growth, and architecture of chestnut-leaved oak (*Quercus castaneifolia*). *iForest Biogeosci. For.* **2016**, *10*, 145–153. [CrossRef]

75. Hathaway, R.L.; Penny, D. Root strength in some *Populus* and *Salix* clones. *N. Z. J. Bot.* **1975**, *13*, 333–344. [CrossRef]

76. Genet, M.; Kokutse, N.; Stokes, A.; Fourcaud, T.; Cai, X.; Ji, J.; Mickovski, S. Root reinforcement in plantations of *Cryptomeria japonica* D. Don: Effect of tree age and stand structure on slope stability. *For. Ecol. Manag.* **2008**, *256*, 1517–1526. [CrossRef]

Article

Comprehensive Analysis of the *TIFY* Gene Family and Its Expression Profiles under Phytohormone Treatment and Abiotic Stresses in Roots of *Populus trichocarpa*

Hanzeng Wang [1], Xue Leng [1], Xuemei Xu [2] and Chenghao Li [1,*]

[1] State Key Laboratory of Tree Genetics and Breeding, Northeast Forestry University, Harbin 150040, China; hzwang0@163.com (H.W.); lengxue19910514@163.com (X.L.)

[2] Library of Northeast Forestry University, Harbin 150040, China; xuemeixu1231@163.com

* Correspondence: chli@nefu.edu.cn

Received: 29 February 2020; Accepted: 9 March 2020; Published: 13 March 2020

Abstract: The *TIFY* gene family is specific to land plants, exerting immense influence on plant growth and development as well as responses to biotic and abiotic stresses. Here, we identify 25 *TIFY* genes in the poplar (*Populus trichocarpa*) genome. Phylogenetic tree analysis revealed these *PtrTIFY* genes were divided into four subfamilies within two groups. Promoter *cis*-element analysis indicated most *PtrTIFY* genes possess stress- and phytohormone-related *cis*-elements. Quantitative real-time reverse transcription polymerase chain reaction (qRT–PCR) analysis showed that *PtrTIFY* genes displayed different expression patterns in roots under abscisic acid, methyl jasmonate, and salicylic acid treatments, and drought, heat, and cold stresses. The protein interaction network indicated that members of the *PtrTIFY* family may interact with COI1, MYC2/3, and NINJA. Our results provide important information and new insights into the evolution and functions of *TIFY* genes in *P. trichocarpa*.

Keywords: *TIFY*; *Populus trichocarpa*; protein interaction network; phytohormone treatment; abiotic stress

---

1. Introduction

The *TIFY* gene family is specific to land plants and was first found in *Arabidopsis* and annotated as a transcription factor [1,2]. This gene family contains a core motif, TIF[F/Y]XG, previously known as ZIM (a zinc-finger protein expressed in inflorescence meristem) [1], and includes 20 members in *Rice*, 18 members in *Arabidopsis*, and 27 members in *Maize* [3–5]. Depending on the conservative domain, the proteins are divided into two groups: Group I containing the C2C2-GATA (CX2CX20CX2C) conservative domain, which is absent in Group II. Generally, the *TIFY* family can be divided into four subfamilies: *TIFY*, *JAZ*, *PPD*, and *ZML* [4]. The *TIFY* subfamily contains only the TIFY domain. The *JAZ* subfamily is named for its conserved Jas motif with 22 amino acids, which contains the TIFY domain and a C-terminal conserved domain, as well as the distinctive motif of SLX2FX2KRX2RX5PY [6,7]. The *PEAPOD* (*PPD*) subfamily was first found by map-based cloning method in *ppd* mutants, which has an exclusive N-terminal PPD domain of around 50 amino acids, a TIFY domain, and a Jas motif lacking the two conserved amino acids "PY" (SLX2FX2KRX2RX5). Finally, the ZML (ZIM-LIKE) subfamily contains a C-terminal GATA-type zinc-finger domain, a CCT (CONSTANS/CO-like/TOC1) domain, and a TIFY domain. The CCT domain is known to play an important role in light signal transduction and takes part in protein–protein interaction [4]. The *JAZ* (Jasmonate ZIM-domain) subfamily has been studied mostly as a key regulator of jasmonate signaling pathways [8]. The CCT domain is important because it allows the proteins of the JAZ subfamily to interact with other proteins to regulate jasmonate signaling and take part in response to biotic and abiotic stresses in several plants [7]. For example, the F-box protein CORONATINE INSENSITIVE1 (COI1) acts as a hormone co-receptor and functions as a negative regulator of the jasmonate signaling pathway [8,9]. JAZ1 and JAZ9 can interact with COI1 in *Arabidopsis* and transgenic plants exhibited JA-insensitive phenotypes and increased resistance to pathogens [10]. Also, JA regulates the ethylene-stabilized transportation factors EIN3 and EIL1 by binding with the JAZ proteins to pathogen defense and development processes [11]. The R2R3 MYB-type transcription factors MYB21 and MYB24 can interact with JAZ1, JAZ8, and JAZ11 in yeast and plants to affect jasmonate-regulated stamen development in *Arabidopsis*. Recently, Ju et al. reported that the C-terminal region of TaJAZ1 interacts with TaABI5 to modulate seed germination and negatively modulate ABA response in wheat [12].

Functional analysis of the *TIFY* gene family has been carried out in several plant species. *PnJAZ1* increased tolerance to salt stress and decreased ABA sensitivity during seed germination and early development in *Pohlia nutans* [13]. As reported, *ZmJAZ14* was significantly induced by polyethylene glycol (PEG), MeJA, abscisic acid (ABA), and gibberellins (GAs). Overexpression of *ZmJAZ14* in Arabidopsis enhanced the plants' tolerance to JA and ABA treatments [14]. Additionally, AsJAZ1 can be involved in nodule development and nitrogen fixation by interacting with AsB2510 in *Astragalus sinicus* [15]. The PPD genes regulate tissue growth, modify lamina size, and restrict curvature of the leaf blade [16]. Removing PEAPOD results in increased leaf lamina size and altered shape of the silique [16]. In *Arabidopsis*, AtPPD2 interacts with AtLHP1 to directly or indirectly regulate several target genes to affect lateral organ development, while *Arabidopsis ppd2* shows greater leaf breadth than wild-type (WT) [17]. In *Maize*, *ZmTIFY4*, *5*, *8*, *26*, and *28* respond to drought; additionally, *ZmTIFY1 19* and *28* show upregulation when treated by three kinds of pathogens [5]. In summary, different *TIFY* subfamily members have different functions during plant growth and development, as well as different responses to phytohormone treatment and abiotic stress.

Populus trichocarpa serves as a model forest species and is widely used in forest tree genomics studies [18]. Previous works identified 24 *PtrTIFY* genes in the *P. trichocarpa* genome and analyzed their expression profiles under JA, MeJA, SA, salt, cold stresses, and pathogen infection in leaves [19,20]. In our study, besides these 24 genes, we found a new *TIFY* family member in the *P. trichocarpa* genome; because of the existence of the Jas and TIFY motif, we named it *PtrJAZ13*. Roots play an important role in woody plants' responses to phytohormone treatment and abiotic stress, but there are no reports to date that focus on the *TIFY* gene family in *P. trichocarpa* roots responding to such manipulations. We therefore used the roots of *P. trichocarpa* as experimental materials and employed quantitative real-time reverse transcription polymerase chain reaction (qRT-PCR) to analyze the expression profiles of all *TIFY* family genes under ABA, MeJA, and SA treatments plus drought, heat, and cold stresses. Our results provide new insights for future investigations into the roles of these candidate genes under hormone treatment and abiotic stress in *P. trichocarpa*.

2. Materials and Methods

2.1. Identification of TIFY Genes in P. trichocarpa

We used the protein family database Pfam (http://pfam.xfam.org/) [21] to search for the hidden Markov model (HMM) profile of the *TIFY* gene family (protein family ID:PF06200) based on an expected value (E-value) cutoff of 0.01 in HMMER 3.3 (http://hmmer.org/) to find *TIFY* genes in the *P. trichocarpa* genome. The genomic library, coding sequence (CDS), and protein database of *P. trichocarpa* were directly downloaded from the Phytozome v12.0 (https://phytozome.jgi.doe.gov) and NCBI (http://www.ncbi.nlm.nih.gov/). The Clustal X (version 2.0) and NCBI database were used to search the identification of the TIFY, ZML, PPD, and JAZ domains. We used the WoLF PSORT database to predict the subcellular localization of *PtrTIFY* genes. Aliphatic index, instability index, and GRAVY (grand average of hydropathy) of *PtrTIFY* genes were identified for all *PtrTIFY* proteins by using ProtParam of Expasy tools (http://web.expasy.org/protparam) [22].

2.2. Phylogenetic Analysis

The ClustalX (version 2.0, http://www.clustal.org/clustal2/) [23] program and Bioedit 7.2 [24] software were used to performed multiple sequence alignment using the full-length protein sequence. In order to analyze the molecular features and phylogenetic relationships of the *TIFY* gene family in plants, MEGA 7.0 [25] was used to build an unrooted phylogenetic tree using the neighbor-joining (NJ) method, with a bootstrap test performed using 1000 replications and the poisson model. Gene clusters referring to the homologs within the three species (*P. trichocarpa, A. thaliana,* and *O. sativa*) which identified based on the NCBI database by BLAST and query cover >75% was regarded as homologous genes. The *PtrTIFY* gene exon/intron organization was determined using GSDS2.0 online software (http://gsds.cbi.pku.edu.cn) [26]. MEME 5.1.0 [27] was used to find motifs in *PuTIFY* genes using the default parameters and a conserved motif number of 15.

2.3. Promoter Cis-Element Analysis

The NCBI and Phytozome V12.0 (https://phytozome.jgi.doe.gov/pz/portal.html) databases were used to search for cis-regulating elements, which were limited the 2000 bp upstream of the ATG codon for each of the analyzed genes [28]. The online software Plantcare (http://bioinformatics.psb.ugent.be/webtools/plantcare/html/) [29] was used to predict and locate their *cis*-elements and analyze the functions of the *TIFY* gene family.

2.4. Plant Materials, Abiotic Stress, and Phytohormone Treatment

For plant samples, the clonally propagated *P. trichocarpa* (genotype Nisqually-1) was grown in woody plant medium (WPM) supplied with 20 g/L sucrose and 5.5 g agar. In vitro plants were cultured in 250 mL plastic containers and grown in a growth chamber with the 16 h light/8 h dark cycling at 23–25 °C with a light intensity of 46 μmol photons m^{-2} s^{-1} irradiation. Two-week-old in vitro plants were used for abiotic stress and phytohormone treatment. For abiotic stress, 7% PEG6000 simulated drought stress, 42 °C simulated heat stress, and 4 °C simulated cold stress. For phytohormone treatments, plants were exposed to a WPM medium containing 100 μM MeJA, 200 μM ABA, or 100 μM SA. Each treatment lasted for 0, 3, 6, 12, and 24 h. The 42 °C stress time points were at 0, 0.5, 1, and 3 h. Untreated in vitro plants at each time point served as control. Fresh roots undergoing each stress were frozen in liquid nitrogen immediately after being harvested and stored at −80°C for later use. Three independent biological replicates were performed for each treatment to ensure reliable data.

2.5. RNA Isolation and Real-Time Quantitative PCR

Highly purified RNA was extracted using the CTAB method [30]. The TransScript® First-Strand cDNA Synthesis SuperMix (TransGen) was used to obtain the cDNA, with the following reaction system: 500 ng of total RNA, 1 μL of anchored oligo$(dT)_{18}$, 10 μL of 2 × ES reaction mix, 1 μL of EasyScript®RT/RI enzyme mix and appropriate ddH_2O to make the total volume to 20 μL. The cDNA was diluted 10-fold (cDNA:nuclease free water = 1:10) for further qRT–PCR analysis. Gene-specific primers were designed using Primer Premier 6.25, and NCBI primer-BLAST was used to check their specificity. TransStart®Top Green qPCR Super Mix (TransGen Biotech, Beijing, China) was employed to carry out the qRT–PCR, with the following reaction system: 10 μL of 2 × TransStart® Top Green qPCR Super Mix, 7 μL of double-distilled H_2O, 1 μL of diluted template, and 2 μL of forward primer and reverse primer. qTOWER 3G Cycler and qPCR software (Analytik Jena, Germany) were used as a work program and the $2^{-\Delta\Delta CT}$ method [31] was used to perform the relative gene expression level analysis. The *PtrActin* was used as the internal reference gene. Triplicate independent technical and biological replications were used in the analysis. The primers for qRT–PCR are listed in Table S1.

2.6. Protein Interaction Network Analysis

The online database STRING v11.0 (https://string-db.org) [32] was used to predict the *PtrTIFY* protein interaction network according to the corresponding homologs between *P. trichocarpa* and *Arabidopsis*. Each amino acid sequence of the *PtrTIFY* genes was used to generate the protein interaction network. In the STRING database, a variety of active interaction sources were provided for the user to select the evidence of interaction [32]. In this study, active interaction sources were selected as "experiments" and "co-expression" with a minimum required interaction score of 0.4. Cytoscape 3.7.2 software [33] was used to visualize and analyze the interactions.

2.7. Statistical Analysis

SPSS software (version 20, IBM, Chicago, USA) was used to analyze data using one-way analysis of variance (ANOVA) followed by Tukey's test to assess significant differences between the control and each treatment. Significance was defined as * $p < 0.05$, ** $p < 0.01$.

3. Results

3.1. Identification of TIFY Genes in P. trichocarpa

In order to identify all assumptive *TIFY* genes in *P. trichocarpa*, the HMM profile of the TIFY domain (protein ID:PF06200) was used as a query to compare with the *P. trichocarpa* genome. In the end, a total of 25 candidate *TIFY* gene members were obtained. Wang et al. found 24 *PtrTIFY* genes in the *P. trichocarpa* genome; in our study, besides the founded 24 *PtrTIFY* genes, we found another *PtrTIFY* gene (*PtrJAZ13*). According to distinct domain, we named TIFY proteins as *PtrTIFY* (1–2), *PtrJAZ* (1–13), *PtrZML* (1–8), and *PtrPPD* (1–2). WoLF PSORT was used to predict the subcellular location of each candidate TIFY protein in *P. trichocarpa*. The results showed that *PtrJAZ6* and *PtrJAZ10* were predicted to be in the cytoplasm and chloroplasts, respectively; *PtrTIFY* (*PtrTIFY1*, -2) and *PtrPPD* (*PtrPPD1*, -2) subfamily members were predicted to be in the nucleus; *PtrZML2*, *-3*, *-4*, *-6* and *-7* were predicted to be in the nucleus, *PtrZML1* and *PtrZML5* were predicted to be in the cytoplasm, and *PtrZML8* was predicted to be in chloroplasts. The *TIFY* gene family has all ten different kinds of TIFY formation in *P. trichocarpa*. *PtrTIFY1*, *-2*, two *PtrJAZ* subfamily members (*PtrJAZ6*, *-12*), and *PtrPPD1*, *-2* shared the TIFY motifs with TIFYGG and TIFYCG, respectively. *PtrJAZ1*, *-4*, *-9*, *-13* contained the TIFYNG motif and *PtrJAZ2*, *-3*, *-5*, *-7*, *-8*, *-10*, *-11* shared the TIFYAG conserved motif. Eight *PtrZML* subfamily members own six different kinds of TIFY formation, including TLSFEG, TLTFRG, TLSFEG, TLSFQG, TIAFEG, TLTFQG. The GRAVY ranged from −0.225 to −0.893, the instability index ranged from 29.88 to 81.91, and the aliphatic index ranged from 48.96 to 85.07. *PtrJAZ13* had the highest stability index; *PtrZML7* possessed the minimum GRAVY (grand average of hydropathy), instability index, and aliphatic index; and *PtrZML5* contained the maximum aliphatic index. Detailed information about *PtrTIFY* genes is given in Table 1.

Table 1. Basic information of the *TIFY* family in *P. trichocarpa*.

Name	Accession Number	Subcellular Location	TIFY Motif	GRAVY	Instability Index	Aliphatic Index
PtrTIFY1	POPTR_0006s26370	nucleus	TIFYGG	−0.686	52.89	57.67
PtrTIFY2	POPTR_0018s01160	nucleus	TIFYGG	−0.642	49.32	55.05
PtrPPD1	POPTR_0002s04920	nucleus	TIFYCG	−0.662	59.33	73.36
PtrPPD2	POPTR_0005s23600	nucleus	TIFYCG	−0.743	60.38	68.33
PtrZML1	POPTR_0002s11140	cytoplasm	TLSFEG	−0.702	48.49	69.33
PtrZML2	POPTR_0002s11150	nucleus	TLTFRG	−0.682	44.88	65.52
PtrZML3	POPTR_0005s19550	nucleus	TLSFEG	−0.626	43.65	67.84
PtrZML4	POPTR_005G152800	nucleus	TLTFRG	−0.698	44.14	61.94
PtrZML5	POPTR_007G116500	cytoplasm	TLSFQG	−0.239	30.39	85.07
PtrZML6	POPTR_0007s03120	nucleus	TIAFEG	−0.852	43.09	58.1
PtrZML7	POPTR_0010s25810	nucleus	TLTFQG	−0.893	29.88	48.96
PtrZML8	POPTR_0017s06970	chloroplast	TIAFEG	−0.753	38.35	62.7
PtrJAZ1	POPTR_0001s13240	nucleus	TIFYNG	−0.563	55.46	69.74
PtrJAZ2	POPTR_0001s16640	nucleus	TIFYAG	−0.603	50.76	66.47
PtrJAZ3	POPTR_0003s06670	nucleus	TIFYAG	−0.635	50.12	63.75
PtrJAZ4	POPTR_0003s16350	nucleus	TIFYNG	−0.567	59.08	72.34
PtrJAZ5	POPTR_0006s14160	nucleus	TIFYAG	−0.417	60.09	65.18
PtrJAZ6	POPTR_0006s23390	cytoplasm	TIFYGG	−0.167	56.63	78.26
PtrJAZ7	POPTR_0008s13290	nucleus	TIFYAG	−0.271	49.13	68.32
PtrJAZ8	POPTR_0010s11850	nucleus	TIFYAG	−0.225	59.27	68.89
PtrJAZ9	POPTR_0011s02260	nucleus	TIFYNG	−0.799	73.46	78.52
PtrJAZ10	POPTR_0012s04220	chloroplast	TIFYAG	−0.329	53.25	68.37
PtrJAZ11	POPTR_0015s04880	nucleus	TIFYAG	−0.296	52.6	67.78
PtrJAZ12	POPTR_0018s08300	nucleus	TIFYGG	−0.559	56.38	62.53
PtrJAZ13	POPTR_006G023300	nucleus	TIFYNG	−0.782	81.91	71.67

3.2. Phylogenetic Analysis of TIFY Genes

To gain insight into the evolutionary relationship between TIFY proteins in *P. trichocarpa*, the 25 full-length TIFY protein sequences from *P. trichocarpa*, 17 proteins from *Rice*, and 18 proteins from *Arabidopsis* were used to build the neighbor-joining phylogenetic tree. MEGA 7.0 software was used to visualize the results. As shown in Figure 1, nine clades were identified in the phylogenetic tree, including JAZ I to V, PPD, TIFY I, II, and ZML. Eight PtrTIFY proteins (PtrZML1, -2, -3, -4, -5, -6, -7, -8) were clustered together in Group I, which contained a GATA zinc-finger domain and a CCT motif based on distinct domain. The remaining 17 PtrTIFY proteins were clustered in Group II. This group contained all PtrJAZ, PtrTIFY, and PtrPPD subfamily members. The *PtrJAZ* subfamily contains the largest numbers of members, which were distributed in three of the four JAZ clades, omitting JAZII. *PtrJAZ13* was in the JAZIV clade and belongs to Group II. From the genetic relationship, *PtrJAZ2, -3, -5, -6, -7, -8, -10, -11, PtrZML1, -3,* PtrTIFY1, *-2* and *PtrPPD1, -2* were clustered with TIFY proteins in *Arabidopsis*. This phenomenon indicates that PtrTIFY proteins are more closely related to those of *Arabidopsis* than to those of *Rice*.

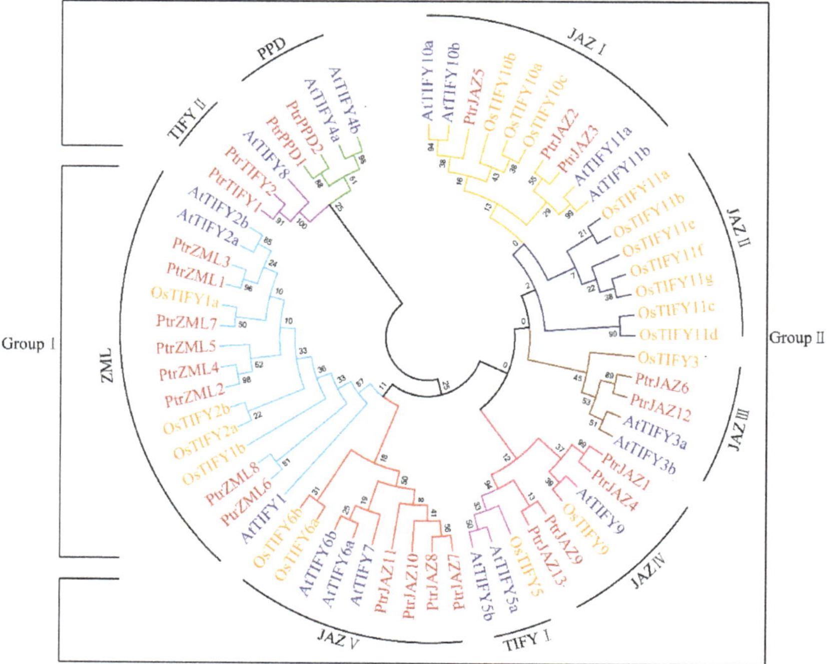

Figure 1. Phylogenetic analysis of the TIFY family members from *P. trichocarpa*, *O. sativa*, and *Arabidopsis*. The phylogenetic trees were constructed using the neighbor-joining (NJ) method of MEGA7 with 1000 bootstrap replicates. For Group I and Group II, Group I contained the C2C2-GATA (CX2CX20CX2C) conservative domain, which was absent in Group II. Each clade was represented by different color lines. The naming approach of each clade was following a previous study [34]. The explanation of group I and II of Figure 1 was added in the Figure 1 legend which was highlighted in red font and yellow background

3.3. Phylogenetic Analysis, Gene Structure, and Conserved Motifs of TIFY Genes in P. trichocarpa

In order to analyze the phylogenetic relationships among the *TIFY* genes in *P. trichocarpa*, an unrooted phylogenic tree was constructed from alignments of the full-length *PtrTIFY* protein sequences. As shown in Figure 2A, *PtrTIFY* genes were clustered into five sections, including two parts of *PtrJAZ*. Based on the phylogenetic analysis, we identified 11 sister pairs, all of which had strong bootstrap support (>90%). To gain further insights into the structural diversity of *PtrTIFY* genes, we analyzed the exon/intron organization in the full-length cDNAs with their corresponding genomic DNA sequences of individual *PtrTIFY* genes in *P. trichocarpa* (Figure 2B). Most *PtrTIFY* subfamily members shared similar exon/intron lengths and numbers, especially the sister pair genes within the same subfamily. For example, both *PtrTIFY1* and *PtrTIFY2* contained five introns of similar lengths, while *PtrJAZ13* and its homologous gene *PtrJAZ9* possessed only one intron. Distinct motifs were found in *PtrTIFY* genes by the MEME website. All *PtrTIFY* genes contained the conserved TIFY motif. As expected, the most closely related genes contained similar motifs, indicating functional diversity among TIFY proteins in the subgroup. In total, we found 15 motifs within 25 *PtrTIFY* genes, including Motif 1 representing the TIFY motif, Motif 2 representing the JAZ motif, and Motif 13 representing the PPD motif. *PtrJAZ9* and *PtrJAZ13* contained the minimum motif in number (Motifs 1 and 2). Interestingly, *PtrJAZ8* possessed one more Motif 6 than *PtrJAZ7* (Figure 2C). Detailed motif information is shown in Table S2.

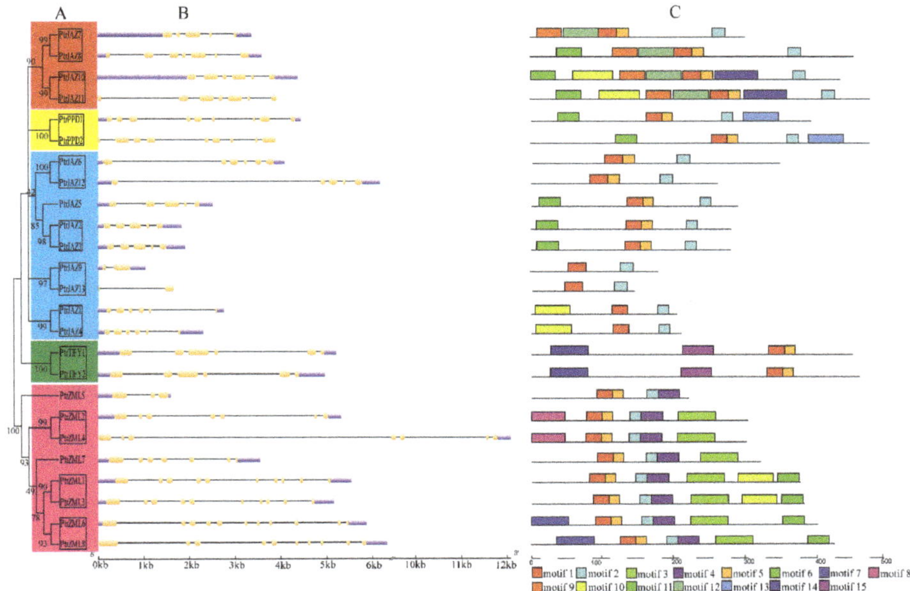

Figure 2. Phylogenetic relationships and gene structure of *TIFYs* in *P. trichocarpa*. (**A**) Multiple alignment of full-length amino acid sequences of *PtrTIFY* genes was carried out with ClustalX 2.0. The phylogenetic tree was constructed using neighbor-joining method with MEGA7.0. The sister pairs was marked with black-box. Red and blue boxes represent the *JAZ* subfamily, yellow box represents the *PPD* subfamily, green box represents the *TIFY* subfamily, and pink box represents the *ZML* subfamily. (**B**) Exon/intron structures of the *PtrTIFY* gene family. (**C**) The conserved motifs were obtained from the MEME website. Fifteen different kinds of conserved motifs were marked with different colors.

3.4. Promoter Cis-Element Analysis

Promoter *cis*-elements play pivotal roles in the transcriptional regulation of genes when plants are under biotic and abiotic conditions. Phytohormones, including ABA, JA, and SA, apart from their functions during growth and development, play important roles in various signal transduction pathways during responses to environmental stresses [35]. Hence, we decided to identify putative *cis*-acting regulatory DNA elements in *PtrTIFY* genes, based on their promoter sequences (2000 bp upstream of the initiation codon). Six different kinds of *cis*-elements relating to phytohormones and environmental stress signals were identified. Most *PtrTIFY* genes possessed the AREB *cis*-elements (cis-acting elements involved in abscisic acid responsiveness). It is worth noting that *PtrJAZ8* contained nine AREB *cis*-elements in different positions of its promoter. *PtrJAZ13*, *PtrZML2*, *-3*, *-5*, and *PtrTIFY2* contained only one AREB *cis*-element. *PtrJAZ4*, *PtrPPD2*, and *PtrZML4* lack the AREB *cis*-element. The remaining *PtrTIFY* genes contained 2–7 AREB *cis*-elements. Thirteen *PtrTIFY* genes contained the CGTCA/TGACG *cis*-element (*cis*-acting element involved in MeJA-responsiveness). *PtrJAZ1*, *-6*, *-7*, *-10*, *-12*, *PtrPPD1*, *-2*, *PtrZML1*, *-2*, *-4*, *-7*, *-8*, and *PtrTIFY2* contained both CGTCA and TGACG *cis*-elements in their promotors. Eight *PtrJAZ* genes, one *PtrPPD* gene, three *PtrZML* genes, and one *PtrTIFY* gene possessed a TCA-element that is involved in SA responsiveness. GAAnnTTC is the typical *cis*-element of heat shock elements (HSE) [36]. In the *PtrTIFY* gene family, *PtrJAZ1*, *-2*, *-3*, *-8*, *-10*, *-13* contained HSE in their promotors, with *PtrJAZ10* possessing the largest number of HSEs. *PtrPPD2*, *PtrZML2*, *6* as well as *PtrTIFY1* and *PtrTIFY2* contained 1–3 HSEs (Figure 3). In summary, each *PtrTIFY* member contained at least one biotic or phytohormone responsive *cis*-element in its promotor. Detailed promoter *cis*-element information is listed in Supplementary Table S3.

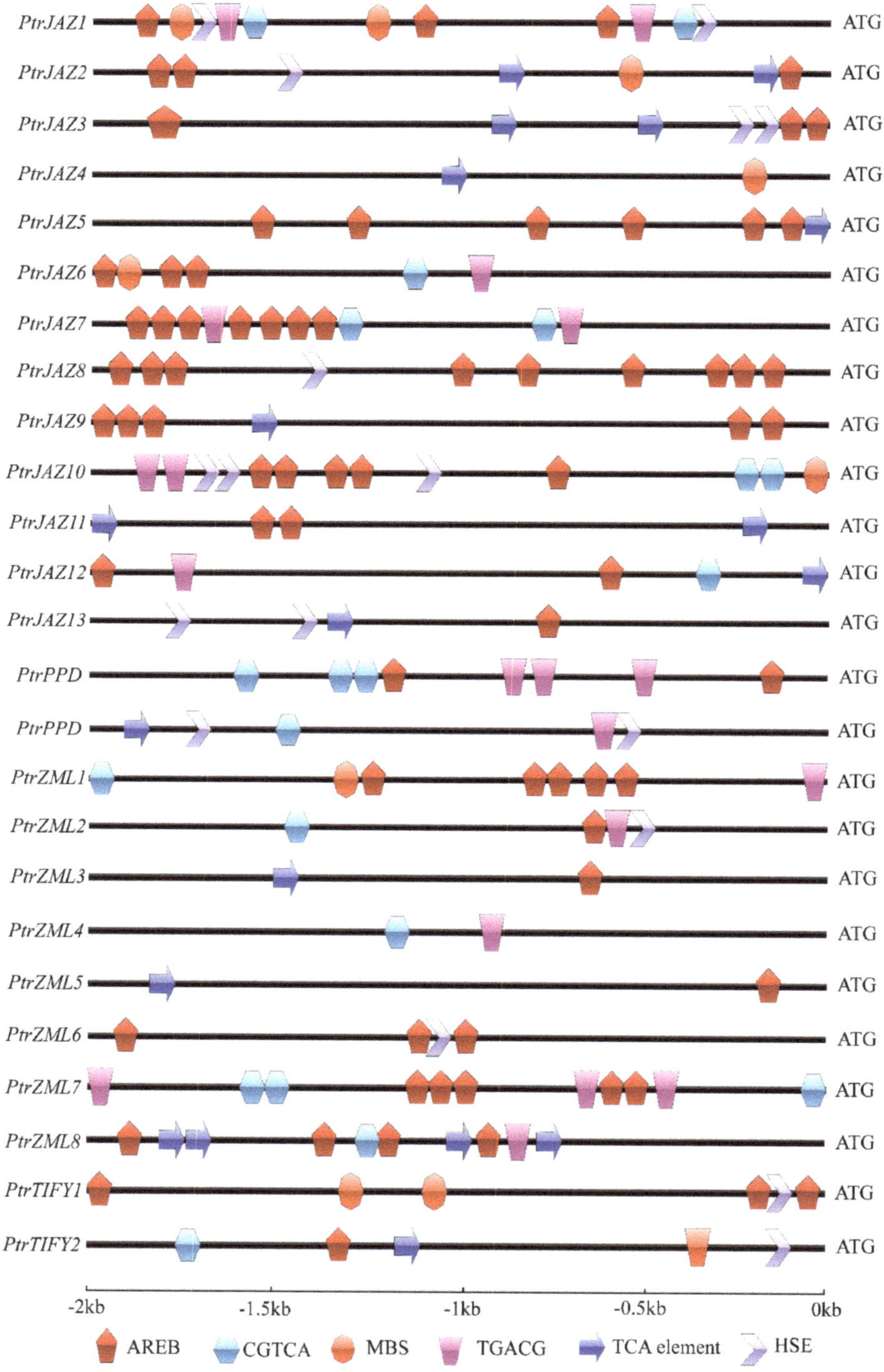

Figure 3. Abiotic stress and phytohormone response elements in *PtrTIFY* gene promoters.

3.5. Expression Pattern of PtrTIFY Genes under Drought, Heat, Cold Stress

To understand how *PtrTIFY* genes take part in drought stress response, we analyzed the expression of *PtrTIFY* genes in roots under drought at 3, 6, 12, and 24 h using qRT–PCR. Our data showed that all *PtrJAZ* subfamily members and *PtrZML2, 3* as well as *PtrZML2* and *PtrZML3* were induced in roots by drought stress. It is noteworthy that *PtrJAZ10* was significantly upregulated at all time points. *PtrPPD1* and *PtrPPD2* as well as *PtrZML5* and *PtrZML8* did not change significantly compared to untreated controls. Finally, *PtrZML4, PtrZML7* and *PtrTIFY2* were downregulated at all time points (Figure 4).

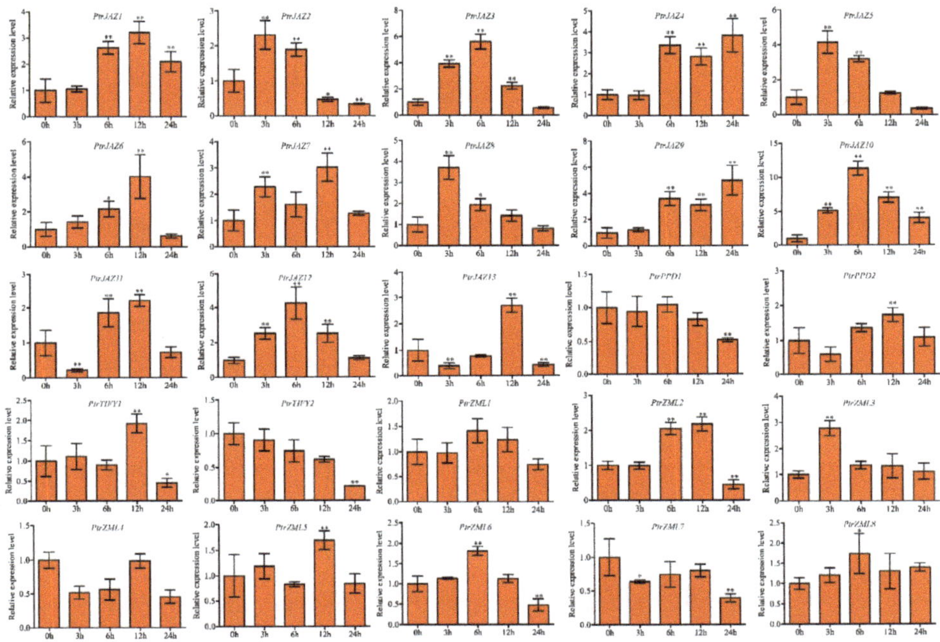

Figure 4. Expression analysis of *PtrTIFY* genes in root under drought stress by qRT–PCR. Error bars represent the standard deviations from three biological replicates. Asterisks indicate stress treatment groups that showed a significant difference in transcript abundance compared with the control group (* $p < 0.05$, ** $p < 0.01$).

We also analyzed the expression pattern of the *PtrTIFY* genes in the roots under 42 °C at 0.5, 1, and 3 h. The results showed all *PtrJAZ* subfamily genes were induced by heat stress except for *PtrJAZ2* and *PtrJAZ13*, which were suppressed at all time points. *PtrJAZ4* was upregulated about 17-fold at 1 h. Interestingly, *PtrJAZ6* was downregulated at the 0.5 h time point and was rapidly upregulated at 1 and 3 h. *PtrJAZ12* responded rapidly to heat stress, with the relative expression level being upregulated 10-fold compared with normal and then gradually declining to 5-fold at 3 h. *PtrPPD2* and *PtrTIFY1* were upregulated at the 1-h time point. *PtrZML3* was highly upregulated at all the time points. Finally, *PtrPPD1, PtrZML1,* and *PtrZML6* showed no change in expression under heat stress at any time point ($p > 0.05$; Figure 5).

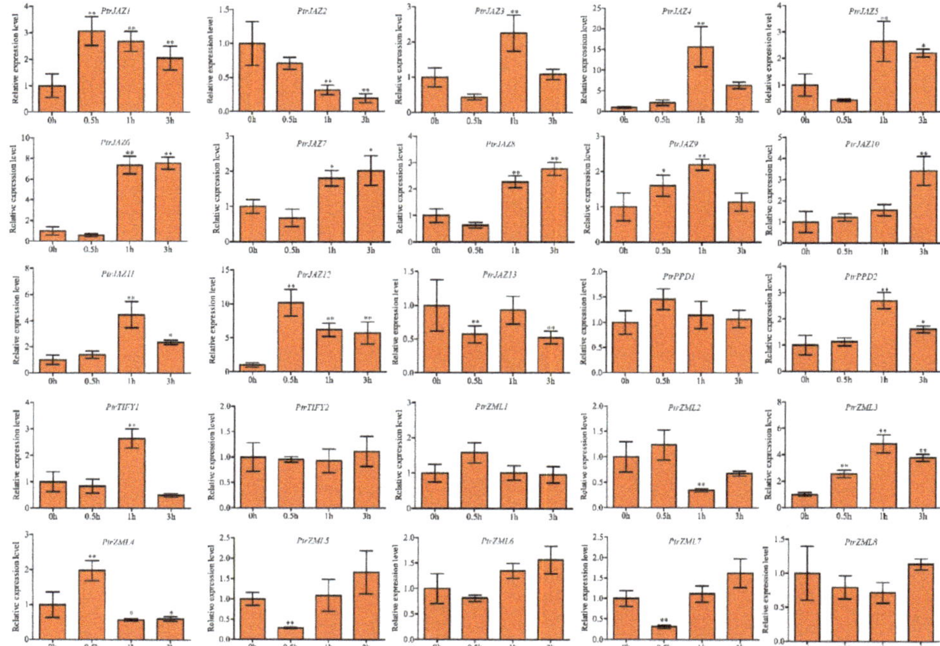

Figure 5. Expression analysis of *PtrTIFY* genes in root under heat stress by qRT–PCR. Error bars represent the standard deviations from three biological replicates. Asterisks indicate stress treatment groups that showed a significant difference in transcript abundance compared with the control group (* $p < 0.05$, ** $p < 0.01$).

For cold stress, eleven *PtrJAZ* subfamily members were induced by cold stress in roots. *PtrJAZ6* was downregulated at 12 h and *PtrJAZ11* was significantly downregulated during the early response period and was elevated back to a level comparable to that of untreated control at 12 and 24 h. *PtrJAZ9* and *PtrJAZ12* were dramatically upregulated at all time points, *PtrJAZ9* and *PtrJAZ12* were upregulated about 11-fold at 24 h and 19-fold at 3 h, respectively. One *PtrPPD* gene and half of the *PtrZML* subfamily genes were strongly induced by cold stress (Figure 6).

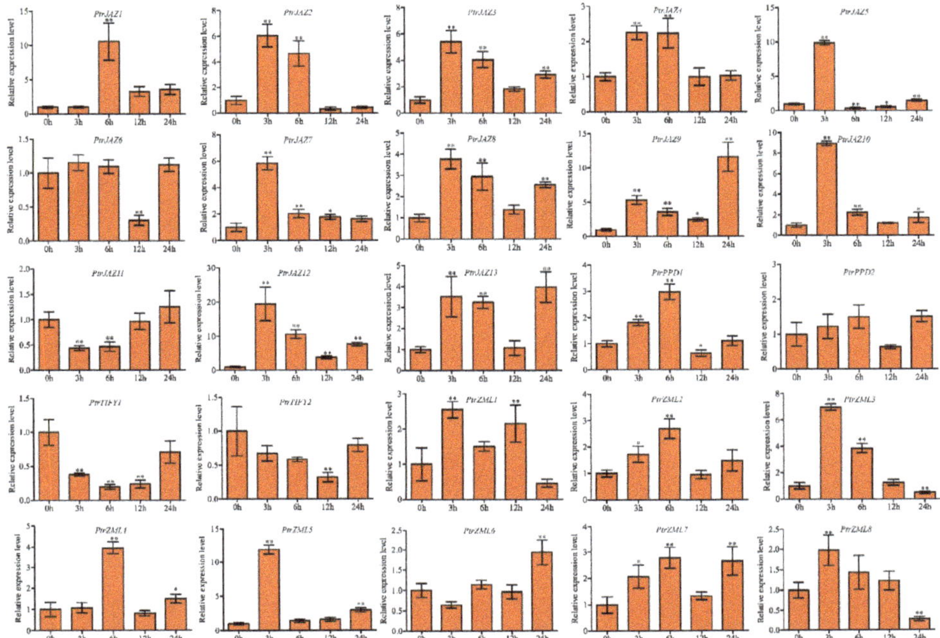

Figure 6. Expression analysis of *PtrTIFY* genes in root under low-temperature stress by qRT–PCR. Error bars represent the standard deviations from three biological replicates. Asterisks indicate stress treatment groups that showed a significant difference in transcript abundance compared with the control group (* $p < 0.05$, ** $p < 0.01$).

3.6. Expression Pattern of PtrTIFY Genes Under ABA, MeJA, and SA Treatments

We employed qRT–PCR to determine the relative expression levels of all *PtrTIFY* members under 200 μM ABA treatment in *P. trichocarpa* roots at 0, 3, 6, 12, and 24 h. Up- or downregulation of the relative expression level >2.0-fold was regarded as significantly differentially expressed. In roots, most *PtrJAZ* subfamily members were induced by 200 μM ABA treatment. The relative expression levels of *PtrJAZ1* and *PtrJAZ9* were upregulated about 17-fold at 12 h compared with control. *PtrJAZ3* and *PtrJAZ6* had similar expression patterns, with the expression level being gradually upregulated at 3 and 6 h and then gradually downregulated at 12 and 24 h. *PtrJAZ13* was downregulated at all time points in the root. *PtrPPD1* and *PtrPPD2* were upregulated by 200 μM ABA treatment at 12 h. *PtrTIFY1* was slightly up-regulated at 3 and 12 h, *PtrTIFY2* was suppressed by ABA at all time points. *PtrZML1*, -5, -6, -7 were significantly upregulated by ABA. The remaining five *PtrZML* subfamily members exhibited opposing expression patterns. The relative expression level of *PtrZML2* showed gradual upregulation at 3 and 6 h, then peaked at 12 h before declining (Figure 7).

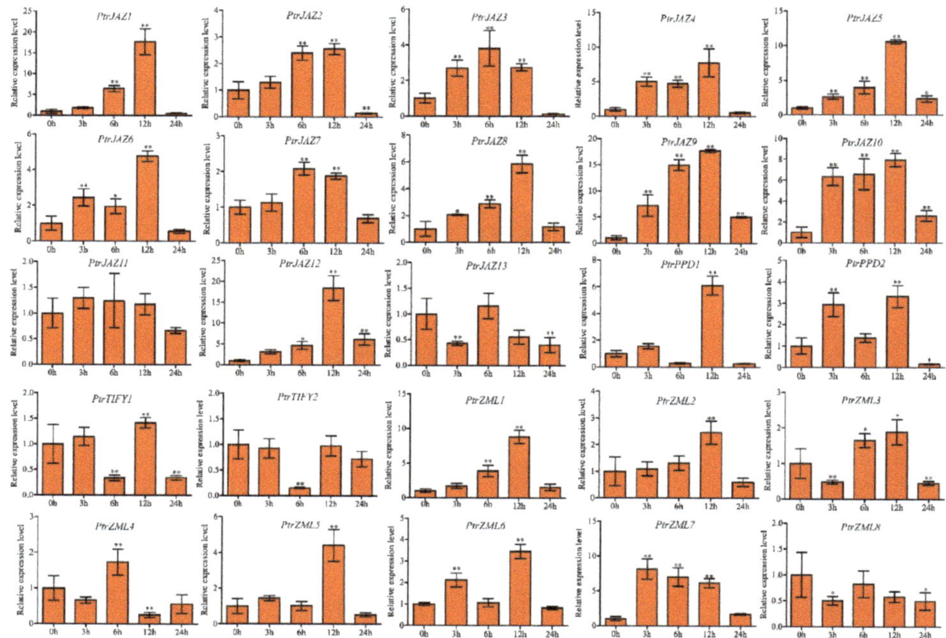

Figure 7. Expression analysis of *PtrTIFY* genes in root under 200 μM ABA treatment by qRT–PCR. Error bars represent the standard deviations from three biological replicates. Asterisks indicate stress treatment groups that showed a significant difference in transcript abundance compared with the control group (* $p < 0.05$, ** $p < 0.01$).

For MeJA treatment, all but *PtrJAZ10* among *PtrJAZ* subfamily members were strongly induced by 100 μM MeJA in roots. *PtrJAZ10* showed no change compared to control. *PtrJAZ4*, *-5*, *-7*, *-11*, *-12*, and *-13* were significantly upregulated at both 3 and 6 h, indicating these *PtrJAZ* subfamily genes are early-responding genes for MeJA treatment. *PtrPPD1* was suppressed by MeJA at 3, 6, and 24 h, while its homologous gene *PtrPPD2* showed no significant change in expression compared to control. *PtrTIFY1* and *PtrTIFY2* were markedly downregulated at 12 h. Only *PtrZML4*, *-7*, and *-8* were induced by MeJA treatment. *PtrZML5* and *PtrZML6* did not respond to MeJA treatment at any time point (Figure 8).

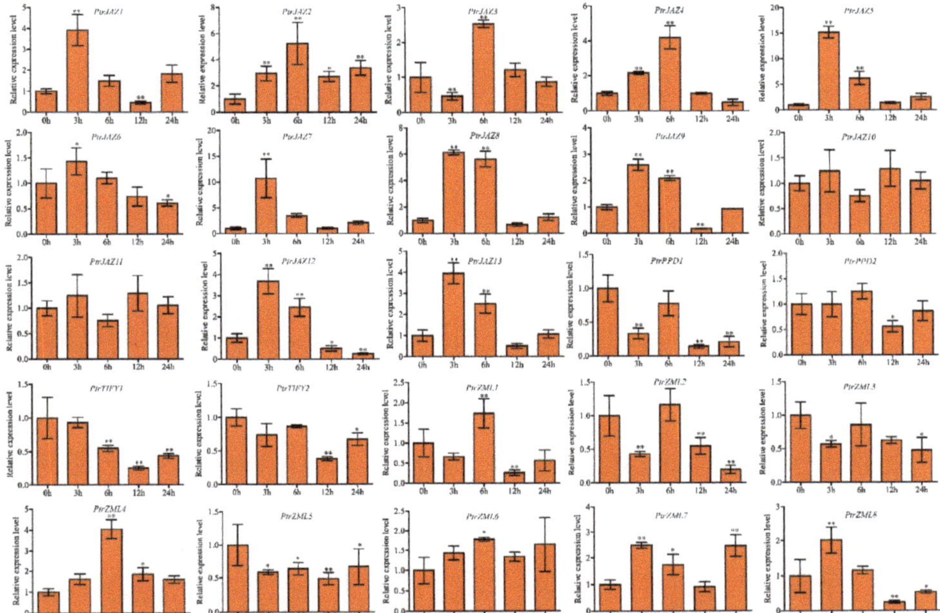

Figure 8. Expression analysis of *PtrTIFY* genes in root under 100 μM MeJA treatment by qRT–PCR. Error bars represent the standard deviations from three biological replicates. Asterisks indicate stress treatment groups that showed a significant difference in transcript abundance compared with the control group (* $p < 0.05$, ** $p < 0.01$).

For SA treatment, all but *PtrJAZ5* were induced by SA treatment in roots. *PtrJAZ1* showed gradual upregulation with time, peaked at 12 h (26-fold) then declined. *PtrJAZ13* was significantly upregulated at 3 and 6 h, then declined to normal expression level at 12 and 24 h. *PtrJAZ1* and *PtrJAZ9* were dramatically upregulated at all time points. *PtrPPD1* was induced by SA treatment in roots and *PtrPPD2* was slightly up-regulated at 3 and 6 h, and repressed at 12 and 24 h. *PtrTIFY2* was suppressed by SA. For *PtrZMLs*, only *PtrZML5* was upregulated at the 6-h time point; *PtrZML1*, *-3*, *-4*, *-6*, *-7*, and *-8* were downregulated compared to the untreated control, while *PtrZML2* was slightly increased by the 100-μM SA treatment. *PtrZML1* and *PtrZML6* were suppressed by MeJA at all the time points (Figure 9).

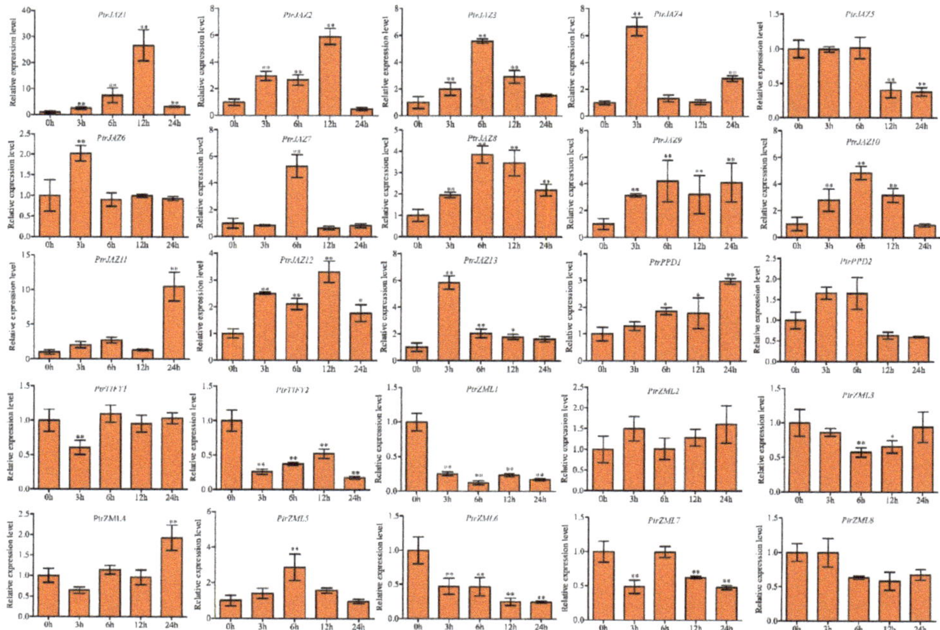

Figure 9. Expression analysis of *PtrTIFY* genes in root under 100 μM SA treatment by qRT–PCR. Error bars represent the standard deviations from three biological replicates. Asterisks indicate stress treatment groups that showed a significant difference in transcript abundance compared with the control group (* $p < 0.05$, ** $p < 0.01$).

3.7. Analysis of the TIFY Protein Interaction Network in P. trichocarpa

The STRING database was used to investigate the relationship of interaction and association in PtrTIFY proteins with a medium confidence score of 0.40 and the interaction sources "experiments" and "co-expression". Thirteen putative protein networks were constructed within three kinds of subfamilies (*PtrJAZ*, *PtrZML*, and *PtrTIFY*). This result showed that PtrTIFY proteins could interact with other proteins to respond to phytohormone treatment and environmental stress in *P. trichocarpa*. Interestingly, the pairs PtrJAZ10 and PtrJAZ11, PtrTIFY1 and PtrTIFY2, and PtrZML5 and PtrZML7 shared similar protein interaction networks, indicating a conserved protein interaction domain may exist in each protein pair (Figure 10).

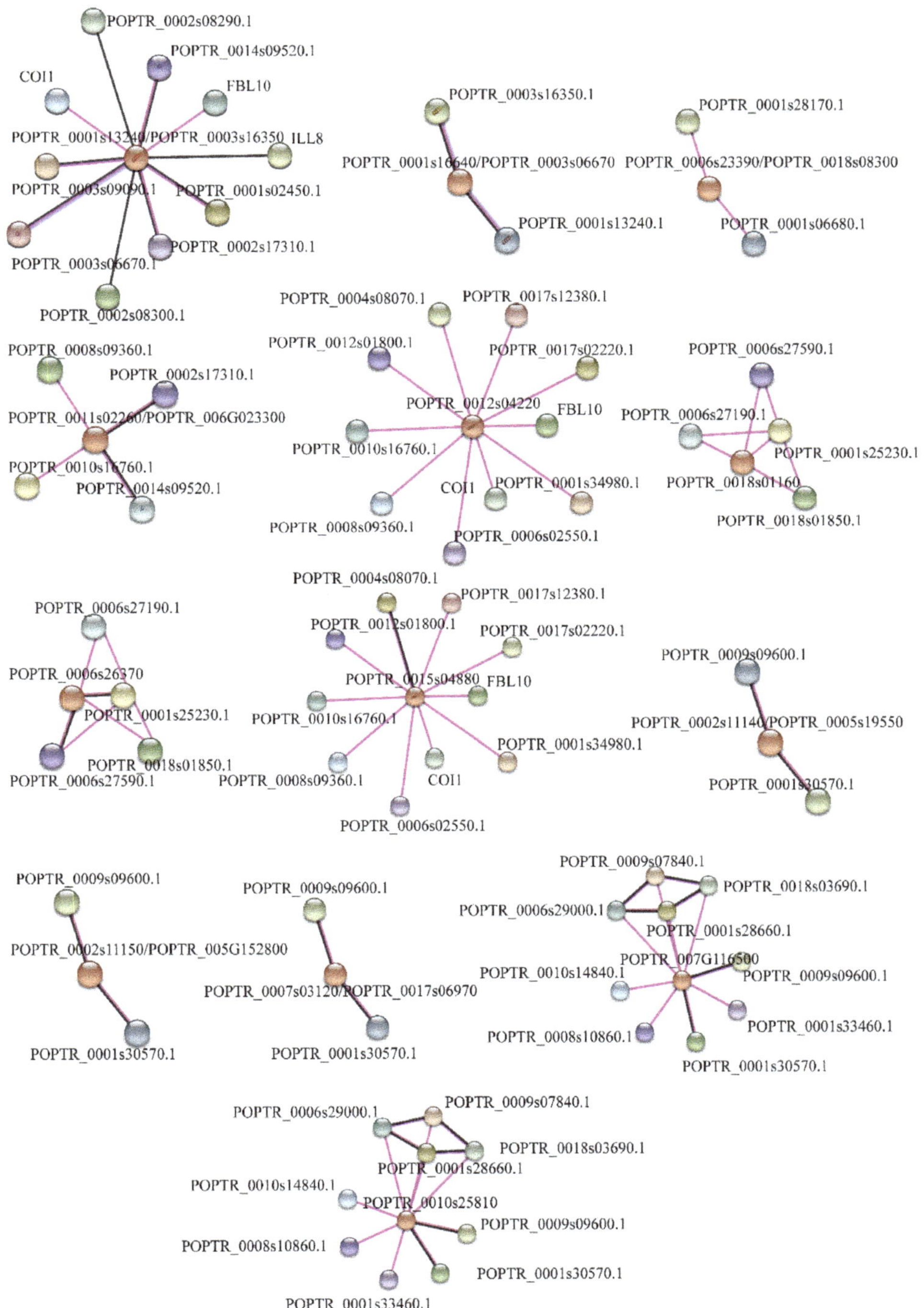

Figure 10. The predicted protein interaction network of PtrTIFY proteins by STRING database. Different colored lines represent different evidence of interaction.

4. Discussion

TIFY is a plant-specific transcription factor that has been identified in many model plants, such as *Arabidopsis*, *Rice*, and *Maize* [1,3,5]. Previous studies have given a comprehensive analysis of 24 *PtrTIFY* genes in leaves of *P. trichocarpa* under various phytohormone treatments and abiotic stresses [19,20]. In this study, we found a new *PtrTIFY* gene, *PtrJAZ13*, that belongs to the important *JAZ* subfamily. Additionally, we examined in roots, for the first time, the expression pattern of the entire *PtrTIFY* family under ABA, MeJA, and SA treatments plus drought, heat, and cold stresses.

Each of the 25 *TIFY* genes in the *P. trichocarpa* genome contains at least one conserved TIFY domain. The length of these sequences varied significantly, implying a high degree of complexity among the *TIFY* genes. WoLF PSORT analysis helped to predict the location of PtrTIFY proteins. Most PtrTIFY proteins (20/25) were predicted to localize to the nucleus. In addition, *PtrZML1*, *-5*, and *PtrJAZ6* on one hand, and *PtrZML8* and *PtrJAZ10* on the other, were localized in cytoplasm and chloroplast, respectively. Interestingly, *PtrJAZ6* and *PtrJAZ12* were in cytoplasm and nucleus, respectively, although they were placed in the JAZIII clade as homologous genes; it is the same with *PtrZML1* and *PtrZML3*, *PtrZML6* and *PtrZML8*. These results indicate that the same phylogenetic group identified based on sequence similarity does not necessarily correspond to the same subcellular localization [28]. Therefore, homologous genes may show differences in gene function and signal transduction.

In this study, we compare the members of the *TIFY* genes in *Populus*, *Rice*, and *Arabidopsis*. Our results show that the phylogenetic tree is divided into two groups; Group I contains the GATA zinc-finger, and Group II lacks it. All *PtrZML* members, together with *OsTIFY1a*, *-1b*, *-2a*, and *-2b*, *AtTIFY1*, *-2a*, and *-2b* belong to the Group I, which contain not only the TIFY domain but also the GATA zinc-finger and CCT motif. The remainder are in Group II. PtrTIFY and AtTIFY proteins are more closely related than those of rice, indicating that they may have a common ancestor. Interestingly, *OsTIFY1a*, *-1b*, *-2a*, *-2b*, *AtTIFY1*, *-2a*, *-2b* not only have a TIFY domain but also a GATA zinc-finger and a CCT motif; the remaining members in *Rice* and *Arabidopsis* do not have the CCT motif. However, almost all the PtrZML proteins contain the CCT motif, suggesting that woody plants might have undergone many different changes during the evolutionary process [37].

The promoter is a specific DNA region located about 2000 bp upstream of the initiation codon that contains a variety of *cis*-elements, including AREB, MBS, LTR, HSE, CGTCA, and TGACG. These help plants respond to exogenous phytohormone treatment and environmental stress [38,39]. In our study, *PtrTIFY* genes possessed at least one stress- or phytohormone-related promoter *cis*-element, indicating *PtrTIFY* genes respond to exogenous phytohormone treatments and environmental stresses. It is worth noting that *PtrJAZ3*, -5 and -8 did not possess the MBS *cis*-element in their promoter regions, although our qRT–PCR data showed these three genes were strongly induced in the root by drought stress; we can thus speculate that they may interact with other genes to respond to drought stress.

Abiotic stresses, such as drought, cold, and heat, along with important phytohormones, always influence plant growth and development in the life cycle, and may even cause fatal damage [40]. To adapt and survive under adverse environments, woody plants have evolved a number of responses, including reprogramming of gene expression [40]. In this study, *PtrJAZ1*, *-3*, *-4*, *-5*, *-7*, *-8*, *-9 -10*, *-12* showed high expression levels during drought, heat, and cold stresses, indicating that the *PtrTIFY* family has important functions in responding to these abiotic stresses. *PtrJAZ2* and *PtrJAZ13* were induced by drought and cold stresses but suppressed by heat stress, whereas *PtrJAZ11* was upregulated in drought and heat stresses but downregulated in cold stress. These results suggest that *PtrTIFY* sets up different physiological and biochemical functions to respond to various environmental challenges. Notably, *OsTIFY9*, a *PtrJAZ1* and *PtrJAZ4* orthologous gene, was strongly induced by drought and cold stresses [3], suggesting the homologous genes have similar functions in different species. A previous study has shown the relative expression level of *VvJAZ11* was downregulated by cold stress [2], but according to our qRT–PCR data, its homologous genes in *P. trichocarpa* (*PtrJAZ7*, *-8*, *-10*) are strongly induced by cold stress. This result suggests that their functions might vary in different plant species. Many previous studies had shown that ABA takes part in response to drought stress [41], with multiple drought stress-responsive genes being induced by ABA [42]. In this study, qRT–PCR analysis showed almost all *PtrJAZ* subfamily members were induced by ABA treatment and by drought stress. In *Rice*, *OsJAZ1* was induced by ABA treatment and drought stress, over-expression of *OsJAZ1* lead to a drought-sensitive phonotype under ABA treatment [43]. *PtrJAZ6* and its homologous gene *PtrJAZ12* were simultaneously induced under ABA treatment and drought stress. *PtrJAZ6/12* was the homologous gene of *OsJAZ1*; this result suggests that *PtrJAZ6/12* may negatively function in drought stress under ABA treatment in *P. trichocarpa*. Salicylic acid (SA) is a key molecule in the signal transduction pathway of abiotic responses [44,45]. In this work, qRT–PCR shows that most *PtrJAZ* subfamily members are induced by SA and by heat stress. *PtrJAZ1*, as well as *PtrJAZ4*, were remarkably upregulated at all time points both by SA treatment and heat stress. A previous study showed that SA functions as a plant growth regulator and alleviates the effects on photosynthesis under heat stress [46]. When plants suffer from heat stress, SA could help to increase proline production through the increase in γ-glutamyl kinase and decrease in proline oxidase activity to help protect photosynthesis from heat stress [47]. *PtrJAZ* subfamily genes may play important roles in the cross-talk between SA and heat stress.

JA is also involved in a wide range of plant growth and developmental processes, including root development [48]. Previous research had shown that JA could regulate lateral root formation in *Arabidopsis* and petunia cuttings [49,50]. Furthermore, *PtrJAZ5* was dramatically upregulated at early response time points (3 and 6 h) under MeJA treatment. *GaJAZ5* was the homologous gene of *PtrJAZ5* in *Gossypium arboretum*; previous research had shown that when *GaJAZ5*-overexpression lines were treated with exogenous MeJA, the lateral root and root hair were higher in number compared to untreated transgenic lines in cotton [51]. OsJAZ1 could regulate root development by interacting with OsMYC2 through JA signaling in *Rice* [52]. In this research, besides *PtrJAZ5, PtrJAZ7, -8, -9, -11, -12, -13* were also dramatically upregulated at early response time points (3 and 6 h); moreover, *PtrJAZ2* was significantly upregulated at all time points under MeJA treatment. These results suggest the *PtrJAZ* subfamily members may take part in root growth and development through the JA signaling pathway.

We constructed the protein interaction network to investigate their interactions and associations in *P. trichocarpa* using the STRING database. The results showed that ten PtrJAZ (PtrJAZ1/4, PtrJAZ2/3, PtrJAZ6/12, PtrJAZ9/13, PtrJAZ10, and PtrJAZ11), two PtrTIFY (PtrTIFY1 and PtrTIFY2), and eight PtrZML (PtrZML1/3, PtrZML2/4, PtrZML5, PtrZML6/8, and PtrZML7) subfamily members were predicted to interact with other proteins. Two members of the PtrPPD subfamily could not interact with any other proteins; nevertheless, *PtrPPD1* was induced by ABA, SA, and cold stress, and *PtrPPD2* was upregulated by ABA at 12 h and heat stress at 1 h, indicating that the two *PtrPPD* subfamily members may function by regulating their target genes to respond to abiotic stress and phytohormone treatment. Increasing evidence supports the idea that Arabidopsis COI 1 (CORONATINE IN SENSITIVE 1), NINJA CNOVEL IN TEACTIOR OF JAZ and two bHLH protein family members (MYC2/MYC3) play important roles in JA signaling and function in the JA pathway, such as root growth inhibition, wound response, and abiotic stress response [10,53,54]. *POPTR_0003s09090* is a homologous gene of *AtMYC2*; moreover, *PtrJAZ1/4* shared a high degree of homology with *AtTIFY9*, which had been reported to interact with AtMYC2 to restrain the activity of transcription factor and then promote the expression of JA-relative genes [55]. Meanwhile, according to our expression pattern, *PtrJAZ1/4* was induced by ABA treatment and drought, heat stresses, so we speculate that *PtrJAZ1/4* may interact with *PtrMYC2* in response to phytohormone treatment and abiotic stress by taking part in JA signaling. PtrJAZ1/4, PtrJAZ10, and PtrJAZ11 shared similar protein interaction networks with PtrZML5 and PtrZML7, PtrJAZ2/3 and PtrZJAZ6/12, PtrTIFY1 and PtrTIFY2, PtrZML2/4 and PtrZML6/8, as well as PtrZML5 and PtrZML7. This phenomenon indicates that a conserved protein interaction domain may exist in these proteins that share the same network.

5. Conclusions

In summary, we identified 25 *PtrTIFY* genes in the *P. trichocarpa* genome and performed comprehensive bioinformatics analyses, including phylogenetic analysis, conserved motif analysis, and promoter *cis*-elements analysis. Importantly, a new TIFY member (*PtrJAZ13*) was found. Then, for the first time, we analyzed the expression pattern of *PtrTIFY* genes in roots under ABA, MeJA, and SA treatments plus drought, heat, and cold stresses. Almost all the *PtrTIFY* genes responded to at least one abiotic stress and phytohormone treatment in the root of *P. trichocarpa*. These results indicate that the *PtrTIFY* genes may play important roles in response to phytohormone treatment and abiotic stress. Our study will help to determine which genes deserve further functional characterization.

Supplementary Materials: The following are available online at http://www.mdpi.com/1999-4907/11/3/315/s1. Table S1: Primer pairs for real-time quantitative PCR. Table S2: Motif sequences of *PtrTIFY* genes. Table S3: Phytohormone- and abiotic stress-related *cis*-elements.

Author Contributions: Conceptualization, H.W., and C.L.; experiment design, C.L.; data curation, W.H., L.X., X.X., and C.L.; contribution of materials/analysis tools, H.W., X.L., X.X., and C.L.; writing—original draft, H.W., and X.L.; writing—review and editing, C.L. All authors have read and agreed to the published version of the manuscript.

Funding: This work was supported by the Fundamental Research Funds for the Central Universities of China (Grant No. 2572016AA01), the 111 Project (Grant No. B16010) and the National Natural Science Foundation of China (Grant No. 31670668).

Acknowledgments: We would like to thank TopEdit (www.topeditsci.com) for its linguistic assistance during the preparation of this manuscript.

Conflicts of Interest: The authors declare no conflict of interest.

References

1. Vanholme, B.; Grunewald, W.; Bateman, A.; Kohchi, T.; Gheysen, G. The tify family previously known as ZIM. *Trends Plant Sci.* **2007**, *12*, 239–244. [CrossRef]
2. Zhang, Y.C.; Gao, M.; Singer, S.D.; Fei, Z.J.; Wang, H.; Wang, X.P. Genome-wide identification and analysis of the TIFY gene family in grape. *PLoS ONE* **2012**, *7*, e0044465. [CrossRef] [PubMed]
3. Ye, H.Y.; Du, H.; Tang, N.; Li, X.H.; Xiong, L.Z. Identification and expression profiling analysis of TIFY family genes involved in stress and phytohormone responses in rice. *Plant Mol. Biol.* **2009**, *71*, 291–305. [CrossRef]
4. Bai, Y.H.; Meng, Y.J.; Huang, D.L.; Qi, Y.H.; Chen, M. Origin and evolutionary analysis of the plant-specific TIFY transcription factor family. *Genomics* **2011**, *98*, 128–136. [CrossRef]
5. Zhang, Z.B.; Li, X.L.; Yu, R.; Han, M.; Wu, Z.Y. Isolation, structural analysis, and expression characteristics of the maize TIFY gene family. *Mol Genet Genomics* **2015**, *290*, 1849–1858. [CrossRef]
6. Katsir, L.; Chung, H.S.; Koo, A.J.K.; Howe, G.A. Jasmonate signaling: A conserved mechanism of hormone sensing. *Curr. Opin Plant Biol.* **2008**, *11*, 428–435. [CrossRef] [PubMed]
7. Chini, A.; Fonseca, S.; Chico, J.M.; Fernandez-Calvo, P.; Solano, R. The ZIM domain mediates homo- and heteromeric interactions between Arabidopsis JAZ proteins. *Plant J.* **2009**, *59*, 77–87. [CrossRef]
8. Thines, B.; Katsir, L.; Melotto, M.; Niu, Y.; Mandaokar, A.; Liu, G.H.; Nomura, K.; He, S.Y.; Howe, G.A.; Browse, J. JAZ repressor proteins are targets of the SCFCO11 complex during jasmonate signalling. *Nature* **2007**, *448*, 661–662. [CrossRef]
9. Xie, D.X.; Feys, B.F.; James, S.; Nieto-Rostro, M.; Turner, J.G. COI1: An Arabidopsis gene required for jasmonate-regulated defense and fertility. *Science* **1998**, *280*, 1091–1094. [CrossRef] [PubMed]
10. Melotto, M.; Mecey, C.; Niu, Y.; Chung, H.S.; Katsir, L.; Yao, J.; Zeng, W.Q.; Thines, B.; Staswick, P.; Browse, J.; et al. A critical role of two positively charged amino acids in the Jas motif of Arabidopsis JAZ proteins in mediating coronatine- and jasmonoyl isoleucine-dependent interactions with the COI1F-box protein. *Plant J.* **2008**, *55*, 979–988. [CrossRef] [PubMed]
11. Guo, H.W.; Ecker, J.R. The ethylene signaling pathway: New insights. *Curr. Opin. Plant Biol.* **2004**, *7*, 40–49. [CrossRef] [PubMed]
12. Ju, L.; Jing, Y.X.; Shi, P.T.; Liu, J.; Chen, J.S.; Yan, J.J.; Chu, J.F.; Chen, K.M.; Sun, J.Q. JAZ proteins modulate seed germination through interaction with ABI5 in bread wheat and Arabidopsis. *New Phytol.* **2019**, *223*, 246–260. [CrossRef] [PubMed]

13. Liu, S.H.; Zhang, P.Y.; Li, C.C.; Xia, G.M. The moss jasmonate ZIM-domain protein PnJAZ1 confers salinity tolerance via crosstalk with the abscisic acid signalling pathway. *Plant Sci.* **2019**, *280*, 1–11. [CrossRef] [PubMed]
14. Wang, L.; Chen, R.M. A maize jasmonate Zim-domain protein, ZmJAZ14, associates with the JA, ABA, and GA signaling pathways in transgenic Arabidopsis. *PLoS ONE* **2015**, *10*, e0128957. [CrossRef]
15. Li, Y.X.; Xu, M.; Wang, N.; Li, Y.G. A JAZ protein in Astragalus sinicus interacts with a leghemoglobin through the TIFY domain and is involved in nodule development and nitrogen fixation. *PLoS ONE* **2015**, *10*, e0139964. [CrossRef] [PubMed]
16. White, D.W.R. PEAPOD regulates lamina size and curvature in Arabidopsis. *Proc. Natl. Acad. Sci. USA* **2006**, *103*, 13238–13243. [CrossRef] [PubMed]
17. Zhu, Y.; Luo, X.; Liu, X.; Wu, W.; Cui, X.; He, Y.; Huang, J. Arabidopsis PEAPODs function with LIKE HETEROCHROMATIN PROTEIN1 to regulate lateral organ growth. *J. Integr. Plant Biol.* **2019**. [CrossRef]
18. Yin, T.; DiFazio, S.P.; Gunter, L.E.; Zhang, X.; Sewell, M.M.; Woolbright, S.A.; Allan, G.J.; Kelleher, C.T.; Douglas, C.J.; Wang, M.; et al. Genome structure and emerging evidence of an incipient sex chromosome in Populus. *Genome Res.* **2008**, *18*, 422–430. [CrossRef]
19. Xia, W.X.; Yu, H.Y.; Cao, P.; Luo, J.; Wang, N. Identification of TIFY family genes and analysis of their expression profiles in response to phytohormone treatments and melampsora larici-populina infection in poplar. *Front Plant Sci.* **2017**, *8*, 493. [CrossRef]
20. Wang, Y.; Pan, F.; Chen, D.M.; Chu, W.Y.; Liu, H.L.; Xiang, Y. Genome-wide identification and analysis of the Populus trichocarpa TIFY gene family. *Plant Physiol. Biochem.* **2017**, *115*, 360–371. [CrossRef]
21. Finn, R.D.; Mistry, J.; Schuster-Bockler, B.; Griffiths-Jones, S.; Hollich, V.; Lassmann, T.; Moxon, S.; Marshall, M.; Khanna, A.; Durbin, R.; et al. Pfam: Clans, web tools and services. *Nucleic Acids Res.* **2006**, *34*, D247–D251. [CrossRef] [PubMed]
22. Wilkins, M.R.; Gasteiger, E.; Bairoch, A.; Sanchez, J.C.; Williams, K.L.; Appel, R.D.; Hochstrasser, D.F. Protein identification and analysis tools in the ExPASy server. *Methods Mol. Biol.* **1999**, *112*, 531–552. [CrossRef] [PubMed]
23. Thompson, J.D.; Gibson, T.J.; Plewniak, F.; Jeanmougin, F.; Higgins, D.G. The CLUSTAL_X windows interface: Flexible strategies for multiple sequence alignment aided by quality analysis tools. *Nucleic Acids Res* **1997**, *25*, 4876–4882. [CrossRef] [PubMed]
24. Hall, T.A. BioEdit: A user-friendly biological sequence alignment editor and analysis program for Windows 95/98/NT. *Nucleic Acids Symposium Series* **1999**, *41*, 95–98.
25. Kumar, S.; Stecher, G.; Tamura, K. MEGA7: Molecular evolutionary genetics analysis version 7.0 for bigger datasets. *Mol. Biol. Evol.* **2016**, *33*, 1870–1874. [CrossRef]
26. Hu, B.; Jin, J.P.; Guo, A.Y.; Zhang, H.; Luo, J.C.; Gao, G. GSDS 2.0: An upgraded gene feature visualization server. *Bioinformatics* **2015**, *31*, 1296–1297. [CrossRef]
27. Bailey, T.L.; Boden, M.; Buske, F.A.; Frith, M.; Grant, C.E.; Clementi, L.; Ren, J.Y.; Li, W.W.; Noble, W.S. MEME SUITE: Tools for motif discovery and searching. *Nucleic Acids Res.* **2009**, *37*, W202–W208. [CrossRef]
28. Liu, Q.G.; Wang, Z.C.; Xu, X.M.; Zhang, H.Z.; Li, C.H. Genome-Wide Analysis of C2H2 Zinc-Finger Family Transcription Factors and Their Responses to Abiotic Stresses in Poplar (Populus trichocarpa). *PLoS ONE* **2015**, *10*, e0134753. [CrossRef]
29. Lescot, M.; Dehais, P.; Thijs, G.; Marchal, K.; Moreau, Y.; Van de Peer, Y.; Rouze, P.; Rombauts, S. PlantCARE, a database of plant cis-acting regulatory elements and a portal to tools for in silico analysis of promoter sequences. *Nucleic Acids Res.* **2002**, *30*, 325–327. [CrossRef]
30. Jaakola, L.; Pirttila, A.M.; Halonen, M.; Hohtola, A. Isolation of high quality RNA from bilberry (*Vaccinium myrtillus* L.) fruit. *Mol. Biotechnol.* **2001**, *19*, 201–203. [CrossRef]
31. Zhang, X.; Wollenweber, B.; Jiang, D.; Liu, F.; Zhao, J. Water deficits and heat shock effects on photosynthesis of a transgenic Arabidopsis thaliana constitutively expressing ABP9, a bZIP transcription factor. *J. Exp. Bot.* **2008**, *59*, 839–848. [CrossRef]
32. Szklarczyk, D.; Gable, A.L.; Lyon, D.; Junge, A.; Wyder, S.; Huerta-Cepas, J.; Simonovic, M.; Doncheva, N.T.; Morris, J.H.; Bork, P.; et al. STRING v11: Protein-protein association networks with increased coverage, supporting functional discovery in genome-wide experimental datasets. *Nucleic Acids Res.* **2019**, *47*, D607–D613. [CrossRef]

33. Otasek, D.; Morris, J.H.; Boucas, J.; Pico, A.R.; Demchak, B. Cytoscape Automation: Empowering workflow-based network analysis. *Genome Biol.* **2019**, *20*, 185. [CrossRef]
34. Yang, Y.X.; Ahammed, G.J.; Wan, C.P.; Liu, H.J.; Chen, R.R.; Zhou, Y. Comprehensive analysis of TIFY transcription factors and their expression profiles under Jasmonic acid and abiotic stresses in Watermelon. *Int. J. Genom.* **2019**, *13*, 6813068. [CrossRef]
35. Wang, L.; Su, H.Y.; Han, L.Y.; Wang, C.Q.; Sun, Y.L.; Liu, F.H. Differential expression profiles of poplar MAP kinase kinases in response to abiotic stresses and plant hormones, and overexpression of PtMKK4 improves the drought tolerance of poplar. *Gene* **2014**, *545*, 141–148. [CrossRef]
36. Sung, D.Y.; Kaplan, F.; Lee, K.J.; Guy, C.L. Acquired tolerance to temperature extremes. *Trends Plant Sci.* **2003**, *8*, 179–187. [CrossRef]
37. Wang, Z.C.; Liu, Q.Q.; Wang, H.Z.; Zhang, H.Z.; Xu, X.M.; Li, C.H.; Yang, C.P. Comprehensive analysis of trihelix genes and their expression under biotic and abiotic stresses in Populus trichocarpa. *Sci. Rep-Uk* **2016**, *6*. [CrossRef]
38. Wang, L.W.; He, M.W.; Guo, S.R.; Zhong, M.; Shu, S.; Sun, J. NaCl stress induces CsSAMs gene expression in Cucumis sativus by mediating the binding of CsGT-3b to the GT-1 element within the CsSAMs promoter. *Planta* **2017**, *245*, 889–908. [CrossRef]
39. Fujita, M.; Fujita, Y.; Noutoshi, Y.; Takahashi, F.; Narusaka, Y.; Yamaguchi-Shinozaki, K.; Shinozaki, K. Crosstalk between abiotic and biotic stress responses: A current view from the points of convergence in the stress signaling networks. *Curr. Opin. Plant Biol.* **2006**, *9*, 436–442. [CrossRef]
40. Debnath, M.; Pandey, M.; Bisen, P.S. An omics approach to understand the plant abiotic stress. *Omics* **2011**, *15*, 739–762. [CrossRef]
41. Fujita, Y.; Fujita, M.; Satoh, R.; Maruyama, K.; Parvez, M.M.; Seki, M.; Hiratsu, K.; Ohme-Takagi, M.; Shinozaki, K.; Yamaguchi-Shinozaki, K. AREB1 is a transcription activator of novel ABRE-dependent ABA signaling that enhances drought stress tolerance in Arabidopsis. *Plant Cell* **2005**, *17*, 3470–3488. [CrossRef]
42. Yamaguchi-Shinozaki, K.; Shinozaki, K. Transcriptional regulatory networks in cellular responses and tolerance to dehydration and cold stresses. *Annu. Rev. Plant. Biol.* **2006**, *57*, 781–803. [CrossRef]
43. Seo, J.S.; Joo, J.; Kim, M.J.; Kim, Y.K.; Nahm, B.H.; Song, S.I.; Cheong, J.J.; Lee, J.S.; Kim, J.K.; Do Choi, Y. OsbHLH148, a basic helix-loop-helix protein, interacts with OsJAZ proteins in a jasmonate signaling pathway leading to drought tolerance in rice. *Plant J.* **2011**, *65*, 907–921. [CrossRef]
44. Maiti, S.; Patro, S.; Pal, A.; Dey, N. Identification of a novel salicylic acid inducible endogenous plant promoter regulating expression of CYR1, a CC-NB-LRR type candidate disease resistance gene in Vigna mungo. *Plant Cell Tiss Org. Cult.* **2015**, *120*, 489–505. [CrossRef]
45. Horvath, E.; Szalai, G.; Janda, T. Induction of abiotic stress tolerance by salicylic acid signaling. *J. Plant Growth Regul.* **2007**, *26*, 290–300. [CrossRef]
46. Nazar, R.; Iqbal, N.; Syeed, S.; Khan, N.A. Salicylic acid alleviates decreases in photosynthesis under salt stress by enhancing nitrogen and sulfur assimilation and antioxidant metabolism differentially in two mungbean cultivars. *J. Plant Physiol.* **2011**, *168*, 807–815. [CrossRef]
47. Khan, M.I.; Iqbal, N.; Masood, A.; Per, T.S.; Khan, N.A. Salicylic acid alleviates adverse effects of heat stress on photosynthesis through changes in proline production and ethylene formation. *Plant Signal Behav.* **2013**, *8*, e26374. [CrossRef]
48. Huang, H.; Liu, B.; Liu, L.Y.; Song, S.S. Jasmonate action in plant growth and development. *J. Exp. Bot.* **2017**, *68*, 1349–1359. [CrossRef]
49. Cai, X.T.; Xu, P.; Zhao, P.X.; Liu, R.; Yu, L.H.; Xiang, C.B. Arabidopsis ERF109 mediates cross-talk between jasmonic acid and auxin biosynthesis during lateral root formation. *Nat. Commun.* **2014**, *5*, 5833. [CrossRef]
50. Lischweski, S.; Muchow, A.; Guthorl, D.; Hause, B. Jasmonates act positively in adventitious root formation in petunia cuttings. *BMC Plant Biol.* **2015**, *15*, 229. [CrossRef]
51. Zhao, G.; Song, Y.; Wang, C.X.; Butt, H.I.; Wang, Q.H.; Zhang, C.J.; Yang, Z.R.; Liu, Z.; Chen, E.Y.; Zhang, X.Y.; et al. Genome-wide identification and functional analysis of the TIFY gene family in response to drought in cotton. *Mol. Genet. Genomics* **2016**, *291*, 2173–2187. [CrossRef] [PubMed]
52. Cai, Q.; Yuan, Z.; Chen, M.J.; Yin, C.S.; Luo, Z.J.; Zhao, X.X.; Liang, W.Q.; Hu, J.P.; Zhang, D.B. Jasmonic acid regulates spikelet development in rice. *Nat. Commun.* **2014**, *5*, 3476. [CrossRef]
53. Breeze, E. Master MYCs: MYC_2, the jasmonate signaling "master switch". *Plant Cell* **2019**, *31*, 9–10. [CrossRef] [PubMed]

54. Acosta, I.F.; Gasperini, D.; Chetelat, A.; Stolz, S.; Santuari, L.; Farmer, E.E. Role of NINJA in root jasmonate signaling. *Proc. Natl. Acad. Sci. USA* **2013**, *110*, 15473–15478. [CrossRef]
55. Chung, H.S.; Howe, G.A. A critical role for the TIFY motif in repression of jasmonate signaling by a stabilized splice variant of the JASMONATE ZIM-domain protein JAZ10 in Arabidopsis. *Plant Cell* **2009**, *21*, 131–145. [CrossRef]

MDPI
St. Alban-Anlage 66
4052 Basel
Switzerland
Tel. +41 61 683 77 34
Fax +41 61 302 89 18
www.mdpi.com

Forests Editorial Office
E-mail: forests@mdpi.com
www.mdpi.com/journal/forests

www.ingramcontent.com/pod-product-compliance
Lightning Source LLC
LaVergne TN
LVHW070855170726
843515LV00005B/651

* 9 7 8 3 0 3 6 5 3 9 6 9 0 *